INVESTIGATING THE PEDAGOGY OF MATHEMATICS

How Do Teachers Develop Their Knowledge?

INVESTIGATING THE PEDAGOGY OF MATHEMATICS
How Do Teachers Develop Their Knowledge?

LIANGHUO FAN
University of Southampton, UK

Imperial College Press

Published by

Imperial College Press
57 Shelton Street
Covent Garden
London WC2H 9HE

Distributed by

World Scientific Publishing Co. Pte. Ltd.
5 Toh Tuck Link, Singapore 596224
USA office: 27 Warren Street, Suite 401-402, Hackensack, NJ 07601
UK office: 57 Shelton Street, Covent Garden, London WC2H 9HE

Library of Congress Cataloging-in-Publication Data
Fan, Lianghuo.
  Investigating the pedagogy of mathematics : how do teachers develop their knowledge? / Lianghuo Fan, University of Southampton, UK.
    pages cm
  Includes bibliographical references and index.
  ISBN 978-1-78326-457-5 (hardcover : alk. paper)
  1. Mathematics teachers--Training of--Research. 2. Teachers--Training of--Research. I. Title. II. Title: How do teachers develop their knowledge?
  QA11.2.F36 2014
  510.71--dc23

2014019239

British Library Cataloguing-in-Publication Data
A catalogue record for this book is available from the British Library.

Copyright © 2014 by Imperial College Press

All rights reserved. This book, or parts thereof, may not be reproduced in any form or by any means, electronic or mechanical, including photocopying, recording or any information storage and retrieval system now known or to be invented, without written permission from the Publisher.

For photocopying of material in this volume, please pay a copying fee through the Copyright Clearance Center, Inc., 222 Rosewood Drive, Danvers, MA 01923, USA. In this case permission to photocopy is not required from the publisher.

Typeset by Stallion Press
Email: enquiries@stallionpress.com

Printed in Singapore

Contents

List of Figures ... ix

List of Tables ... xi

Focusing on the Growth of Teachers' Knowledge: An Introduction ... xix

Part I The Chicago Study ... 1

Chapter 1   Introduction ... 3

    Background of the Chicago Study ... 3
    Need for the Study ... 4
    Statement of the Problem ... 5
    Structure of Part I ... 6

Chapter 2   Review of the Literature ... 9

    What is Knowledge? ... 10
    What Knowledge do Teachers Need? ... 16
    What Knowledge do Teachers Have? ... 25
    How do Teachers Develop their Knowledge? ... 29
    Summary of the Review ... 36

Chapter 3   A Conceptual Framework of the Study ... 37

    Knowledge ... 37
    Teachers' Knowledge ... 42

	Teachers' Pedagogical Knowledge	44
	Sources of Teachers' Pedagogical Knowledge . . .	49
	Summary	54
Chapter 4	Research Design and Procedures	57
	Population and Sample	57
	Instruments	59
	Data Collection	64
	Data Processing and Analysis	68
	Strengths and Limitations of the Methodology	71
	Summary of the Methodology	72
Chapter 5	Findings of the Chicago Study (I): Pedagogical Curricular Knowledge	75
	Knowledge of Teaching Materials	76
	Knowledge of Technology	92
	Knowledge of Other Teaching Resources	106
	Summary of the Findings	116
Chapter 6	Findings of the Chicago Study (II): Pedagogical Content Knowledge	119
	Analysis of the Questionnaire Data	121
	Analysis of the Interview Data	130
	Summary of the Findings	152
Chapter 7	Findings of the Chicago Study (III): Pedagogical Instructional Knowledge	155
	Analysis of the Questionnaire Data	156
	Analysis of the Interview Data	160
	Summary of the Findings	183
Chapter 8	Findings of the Chicago Study (IV): Some Other Issues	185
	How Teachers Use Different Sources	186

	How Teachers Improve their Pedagogical Knowledge	201
	Summary of the Findings	205
Chapter 9	Conclusions, Implications, and Recommendations	209
	Summary and Conclusions	209
	Implications for Teacher Educators, School Administrators, and Teachers	213
	Recommendations for Further Study	216
Part II	**The Singapore Study**	**219**
Chapter 10	The Singapore Study	221
	Background and Introduction of the Singapore Study	221
	The Conceptual Framework	226
	Methodological Matters	230
	Findings of the Singapore Study	235
	Summary of the Singapore Study	257
Chapter 11	Comparison and Conclusion	261
	Comparison of the Chicago and Singapore Studies	261
	Concluding Remarks	269
Appendix 1A	Teacher Questionnaire (Chicago Study)	275
Appendix 1B	Notes for Classroom Observation	287
Appendix 1C	Script for Interviewing Teachers	289
Appendix 1D	Script for Interviewing Math Chairs	293
Appendix 1E	A Profile of Teacher Participants (Chicago Study)	295
Appendix 1F	Main Results of Logistic Regression Analyses (Chicago Study)	297
	Pedagogical Curricular Knowledge — Knowledge of Textbooks	297

	Pedagogical Curricular Knowledge — Knowledge of Technology	300
	Pedagogical Curricular Knowledge — Knowledge of Concrete Materials	301
	Pedagogical Instructional Knowledge	302
	Knowledge Sources Used to Represent New Topics	303
	Pedagogical Content Knowledge	303
Appendix 2A	A Profile of Teacher Participants (Singapore Study)	305
Appendix 2B	Teacher Questionnaire (Singapore Study)	307
References		321
Author Index		333
Subject Index		337

List of Figures

2.1	Three types of knowledge for mathematics teachers by Lappan and Theule-Lubienski (1994)	20
3.1	A framework to investigate the sources of teachers' pedagogical knowledge	51
5.1	Knowledge of textbooks being used	77
5.2	Comparison of the contribution of different sources to teachers' knowledge of textbooks	79
5.3	Comparison of the three groups of teachers' average evaluation of the contribution of different sources to their knowledge of textbooks	81
5.4	Comparison of the contribution of different sources to teachers' knowledge of technology	95
5.5	Comparison of the three groups of teachers' average evaluation of the contribution of different sources to teachers' knowledge of technology	97
5.6	Comparison of the contribution of different sources to teachers' knowledge of concrete materials	108
5.7	Comparison of the three groups of teachers' average evaluation of the contribution of different sources to teachers' knowledge of concrete materials	110
6.1	Proportions of teachers who got their PCnK about various topics from different sources	126
7.1	Comparison of the contribution of different sources to teachers' PIK	158

7.2	Comparison of the three groups of teachers' average evaluations of the contribution of different sources to their PIK	160
8.1	Percentages of teachers who were taught different pedagogical knowledge in their preservice training	190
10.1	A framework to investigate the sources of Singaporean teachers' pedagogical knowledge	228
10.2	Percentage of the Singaporean teachers who got their PCnK about various topics they listed from different sources	250

List of Tables

4.1	Distribution of the questions in the questionnaire in terms of the components of teachers' pedagogical knowledge and the knowledge sources	60
4.2	Numbers of students, teachers, and mathematics teachers in the three sample schools (1996–1997)	65
4.3	Response rate of the questionnaire survey	65
4.4	Selection of teachers for classroom observation	66
4.5	Classes observed for the study	67
5.1	Courses taught by teachers	77
5.2	Distributions of the numbers of teachers giving different evaluations about the contribution of various sources to the development of their knowledge of textbooks	78
5.3	Distributions of the numbers of teachers among the three groups giving different evaluations of the contribution of each source to their knowledge of textbooks	81
5.4	Sources of teachers' knowledge of textbooks (from interview data)	92
5.5	How often teachers used calculators and computers in their mathematics classes	93
5.6	Distributions of the numbers of teachers giving different evaluation about the contribution of various sources to the development of their knowledge of how to use technology for teaching mathematics	94

5.7	Distributions of the numbers of teachers among the three groups giving different evaluations of the contribution of each source to their knowledge of technology	97
5.8	Sources of teachers' knowledge of technology (from interview data)	106
5.9	How often teachers used concrete materials to teach mathematics	107
5.10	Distributions of the numbers of teachers giving different evaluations about the contribution of various sources to their knowledge of concrete materials	107
5.11	Distributions of the numbers of teachers among the three groups giving different evaluations of the contribution of each source to their knowledge of concrete materials	112
5.12	Sources of teachers' knowledge of concrete materials (from the interview data)	115
5.13	A summary of the importance of different sources to the development of teachers' PCrK	117
6.1	Distributions of the mathematics topics given by teachers in Questions 20 and 21	122
6.2	Sources of teachers' PCnK reported in the questionnaire	126
6.3	Numbers of teachers who answered the questions among the three groups	128
6.4	Teachers' responses across the three groups about the sources of their PCnK of the topics given in Questions 20 and 21	128
6.5	Teachers' responses across the three groups to the questions of their developing their PCnK from Source E	129
6.6	Sources of teachers' PCnK (from interview data)	151
7.1	Distributions of the numbers of teachers giving different evaluations about the contributions of various sources to their PIK	157

7.2	Distributions of the numbers of teachers among the three groups giving different evaluations of the contribution of each source to their PIK	161
7.3	Teaching strategies and the knowledge sources (TA1)	167
7.4	Teaching strategies and the knowledge sources (TA2)	168
7.5	Teaching strategies and the knowledge sources (TA3)	170
7.6	Teaching strategies and the knowledge sources (TB1)	172
7.7	Teaching strategies and the knowledge sources (TB2)	174
7.8	Teaching strategies and the knowledge sources (TB3)	176
7.9	Teaching strategies and the knowledge sources (TC1)	177
7.10	Teaching strategies and the knowledge sources (TC2)	179
7.11	Teaching strategies and the knowledge sources (TC3)	181
7.12	Sources of teachers' PIK (from interview data)	182
8.1	Teaching strategies used as teachers and encountered as students, reported by teachers	187
8.2	Teachers' evaluation of the usefulness of preservice training courses in enhancing their pedagogical knowledge	191
8.3	Courses taken by teachers in non-degree professional training	193
8.4	Distribution of the majors teachers enrolled in inservice degree programs	193
8.5	Teachers' evaluation of the usefulness of inservice training programs in enhancing their pedagogical knowledge	194

8.6	Teachers' evaluation of the usefulness of other professional training in enhancing their pedagogical knowledge	195
8.7	Teachers' evaluation of the usefulness of organized professional activities at the local/state and regional/national levels	196
8.8	Teachers' evaluation of the usefulness of organized professional activities at the school/department and district/county levels	197
8.9	Frequencies of teachers having non-organized professional activities	198
8.10	Teachers' evaluation of the usefulness of non-organized professional activities in enhancing their pedagogical knowledge	200
8.11	How teachers use different sources to design the ways to represent new mathematics topics	200
9.1	A summary of the main findings on the relative importance of different sources to the development of teachers' pedagogical knowledge	213
10.1	Singaporean students' performances in TIMSS mathematics studies from 1995 to 2011	224
10.2	Numbers of students, teachers, and mathematics teachers in the six Singaporean schools	230
10.3	Selection of teachers for classroom observation in the Singapore study	232
10.4	Classes observed in the Singapore study	233
10.5	Results on how Singaporean teachers gave different evaluations on the contribution of various sources to the development of their knowledge of textbooks	236
10.6	Results of logistic regression on the data about the contribution of different sources to the Singaporean teachers' knowledge of textbooks	237

10.7	Distributions of the numbers of the Singaporean teachers among the three groups giving different evaluations of the contribution of the source "experience as school students" to their knowledge of textbooks	239
10.8	Sources of Singaporean teachers' knowledge of textbooks identified during the interview	240
10.9	Results on how Singaporean teachers gave different evaluations on the contribution of various sources to the development of teachers' knowledge of using technology for teaching mathematics	242
10.10	Results of logistic regression on the data about the contribution of different sources to Singaporean teachers' knowledge of technology	243
10.11	Sources of Singaporean teachers' knowledge of technology identified during the interview	245
10.12	Results on how Singaporean teachers gave different evaluations on the contribution of various sources to the development of their knowledge of using concrete materials/physical models for teaching mathematics	246
10.13	Results of logistic regression on the data about the contribution of different sources to Singaporean teachers' knowledge of concrete materials	247
10.14	Sources of Singaporean teachers' knowledge of concrete materials identified during the interview	248
10.15	Results of logistic regression on the data about the contribution of different sources to the Singaporean teachers' PCnK (binary logistic model)	250
10.16	Singaporean teachers' responses across the three groups to the questions of their developing PCnK from the source of "experience as student"	251
10.17	Singaporean teachers' responses across the three groups to the questions of their developing PCnK from the source of "preservice training"	252

10.18	Sources of Singaporean teachers' PCnK identified during the interview	253
10.19	Results on how Singaporean teachers gave different evaluations on the contribution of various sources to the development of their PIK	254
10.20	Results of logistic regression on the data about the contribution of different sources to the Singaporean teachers' PIK	255
10.21	Sources of Singaporean teachers' PIK identified during the interview	256
10.22	A summary of the main findings on the relative importance of different sources to the development of Singaporean teachers' pedagogical knowledge	258
11.1	Comparison of the contribution of different sources to the development of teachers' curricular knowledge in the Chicago and Singapore studies	262
11.2	Comparison of teachers' average evaluation of the contribution of different sources to teachers' PCrK in the two studies	264
11.3	Comparison of the relative frequency (%) of teachers identifying different sources for their PCnK about the teaching of specific mathematics contents	265
11.4	Comparison of teachers' evaluations of the contribution of different sources to teachers' PCnK	265
11.5	Comparison of the contribution of different sources to teachers' PIK in Chicago and Singapore	267
11.6	Comparison of teachers' average evaluation of the contribution of different sources to teachers' PIK in Chicago and Singapore	268
11.7	Differences about the importance of each source to the development of pedagogical knowledge among the teachers of three groups in Chicago (C) and Singapore (S)	269
11.8	An overview of the results of the Chicago (CHA) and Singapore (SING) studies	270

List of Tables

1E.1	Distribution of teachers by gender and school	295
1E.2	Distribution of teachers by age and school	295
1E.3	Distribution of teachers by highest educational degree and school	296
1E.4	Distribution of the majors and minors of the degrees teachers earned	296
1E.5	Distribution of teachers with different lengths of experiences of teaching mathematics	296
1F.1	Logistic regression on the data about the contribution of different sources to teachers' knowledge of textbooks	298
1F.2	Logistic regression on the data about the contribution of different sources to teachers' knowledge of technology	300
1F.3	Logistic regression on the data about the contribution of different sources to teachers' knowledge of concrete materials	301
1F.4	Logistic regression on the data about the contribution of different sources to teachers' PIK	302
1F.5	Logistic regression on the data about the frequencies of different sources teachers used to design the way to represent new mathematics topics	303
1F.6	Logistic regression on the data about the contribution of different sources to teachers' PCnK	303
2A.1	Numbers of full-time and contract teachers in the six secondary schools in Singapore	305
2A.2	Numbers of male and female teachers in the six secondary schools in Singapore	305
2A.3	Distribution of the teachers by age in the six secondary schools in Singapore	306
2A.4	Academic qualifications of the teachers in the six secondary schools in Singapore	306
2A.5	Distribution of the teachers according to their years of teaching experience in the six secondary schools in Singapore	306

Focusing on the Growth of Teachers' Knowledge: An Introduction

There is no doubt that, over the last few decades, general politicians, educational leaders, researchers, and practitioners have increasingly realized the crucial importance of teachers' knowledge in improving the quality of classroom teaching and learning. Accordingly, the issue of how teachers gain or develop their knowledge has also been a hot topic in educational research and on-going debate. Answering this issue calls for research-based evidences and sound knowledge for policy making and school practice in relation to teacher recruitment, teacher education, and teacher professional development.

This book presents two research studies focusing on how teachers develop their knowledge in the pedagogy of mathematics. The first study was conducted in the city of Chicago of the United States (hereafter called "the Chicago study") and the second one was conducted in Singapore (hereafter called "the Singapore study"). Both studies address the same research questions, but in different social, cultural, and educational contexts. The first question is, are there different sources for teachers to gain their pedagogical knowledge? And the second is, if the answer to the first question is "yes", then how do different sources contribute to the development of teachers' pedagogical knowledge?

The book consists of two main parts. Part I is devoted to the Chicago study in a relatively detailed way. It contains the following nine chapters.

Chapter 1 introduces the background, the significance, and the research questions of the study.

Chapter 2 first presents some epistemological background about the concept of knowledge for the study, and then provides a relatively broad review of the literature in the field of teacher knowledge, which is divided around three major issues: what knowledge do teachers need, what knowledge do teachers have, and how do teachers develop their knowledge.

Chapter 3 establishes a conceptual framework to examine teachers' pedagogical knowledge and the possible different sources from which teachers develop their pedagogical knowledge.

Under the conceptual framework, teachers' pedagogical knowledge is further classified into three components, namely, pedagogical curricular knowledge (PCrK): knowledge of teaching materials and resources, including technology; pedagogical content knowledge (PCnK): knowledge of ways to represent mathematics concepts and procedures; and pedagogical instructional knowledge (PIK): knowledge of general teaching strategies and classroom organizational models.

Chapter 4 introduces the research design, instruments, and procedures used for collecting and analyzing the data, which is gathered through a questionnaire survey, classroom observation, and interviews from 77 mathematics teachers in three high-performing high schools, a stratified random sample from the 25 best high schools in the metropolitan area of Chicago.

Chapters 5, 6, and 7 report the core findings of the Chicago study. More specifically, Chapter 5 reports the results about how teachers develop their PCrK, Chapter 6 is about how teachers develop their PCnK, and Chapter 7 is about how teachers develop their PIK.

Chapter 8 reports other findings of the study, supplementing the three earlier chapters by looking at some other issues concerning the development of teachers' pedagogical knowledge. It looks at each specific source identified in the study and does not separate the general pedagogical knowledge into three different components.

Chapter 9 provides a summary of the Chicago study and its main conclusions. It also discusses the implications of the findings for teacher educators, school administrators, and teachers themselves on how to effectively pursue the growth of teachers' pedagogical knowledge. The chapter also offers suggestions for further research concerning the development of teacher knowledge.

Part II of the book comprises Chapter 10 and Chapter 11. Chapter 10 presents the Singapore study, but in a more concise manner, given the fact that it is largely a replication of the Chicago study in terms of conceptualization and research methods. Accordingly, more emphasis of the writing is placed on the Singapore study's different social, cultural, and educational context compared with the Chicago study's. The Singapore study has a purpose of confirmation and comparison of the findings with the Chicago study. The data was collected from 73 mathematics teachers in six secondary schools, a stratified random sample from all of the 152 secondary schools in Singapore.

Chapter 11 juxtaposes the Singapore study with the Chicago one and compares the similarities and differences of the results as revealed in the two studies. Some concluding remarks are offered at the end of the chapter.

Both the Chicago and Singapore studies revealed that there are various sources from which teachers can develop their pedagogical knowledge. In particular, both studies consistently found that teachers' own teaching experience and reflection is the most important source for them to develop all the three components of pedagogical knowledge, and teachers' daily or informal exchanges with their colleagues is the second most important source for all the three components of their pedagogical knowledge. For the other sources, the studies revealed, to a greater or lesser degree, there exists different importance in teachers' developing one or more components of their pedagogical knowledge. In addition, the teachers in Chicago reported that preservice training was the least important source, while their Singapore counterparts viewed it as a secondarily important source, for the development of their pedagogical knowledge.

It should be pointed out that the Chicago study was originally completed as the author's doctoral thesis and, after translation into Chinese, first published in a book series by East China Normal University Press, Shanghai, China in 2003. As a researcher and the author, I was heartened to know that the book has had a positive impact on China's national policy about school-based teacher professional development,[1] and was regarded

---

[1] See Jian Liu's comments on the Chinese edition of the Chicago study. *A Study on the Development of Teachers' Pedagogical Knowledge* (2nd edition, back cover), Shanghai: East China Normal University Press, 2013.

by researchers, generously, as one of the most influential studies in this area,[2] being widely used as a university graduate course book. According to the information provided by Dr. Dongchen Zhao, it has been cited 833 times in China (source: www.cnki.net, as of 1 March 2014) including 100 doctoral theses but excluding citations in books. In consideration of the demand from the readers, the publisher has decided to publish the second edition of the book in Chinese. I am delighted that the English version of the Chicago study, of which much original integrity is kept with some necessary changes, together with a follow-up study, the Singapore study, is now published by Imperial College Press.

Although the studies presented in this book were conducted in Chicago and Singapore with focus on pedagogical knowledge of mathematics teachers, I hope that the book can be of value not only to readers in these two countries and mathematics educators who will naturally feel it easier to understand the relevant background of the studies, but also to any educational policy makers and researchers, teacher educators, school administrators and teachers, and graduate students who are interested in the theme of teachers' knowledge development. The reason is that the issue of how teachers gain their knowledge has general importance and communality in different countries and across different subject matters, and it is also internationally a common challenge to continuously improve teachers' professional knowledge.

Many people have offered me a lot of important help in conducting the Chicago and Singapore studies and in writing and publishing this book. For this, I am indebted to my professors, colleagues, research assistants, and friends, in particular, Zalman Usiskin, Robert Dreeben, John Craig, Kenneth Wong, Susan Stodolsky, John McConnell, Steve Victora, Lilane Koehn, Timothy Kanold, Kwai Ming Wa, David Witonsky, Alfred Estberg, Suzanne Levin, Eileen Fernández, Susan Chang, and Tad Waddington for the Chicago study, and Christina Cheong, Lee Peng Yee, Lionel Pereira-Mendoza, Kho Tek Hong, Cheong Peck Yoke, Teo Yin Nah, Yeo Jia Chyang, Yeo Jinguang Mathew, and Zhu Yan for the Singapore study.

---

[2]For example, see Zhang (2009).

I am also indebted to all the school teachers, heads of mathematics departments, and administrators in Chicago and Singapore who participated or helped in various ways in the two studies, though because of the agreement for conducting the studies, their names cannot be listed in this book.

I must thank my wife Ellen and our daughters Jenny and Kate for their patience and sacrifice for supporting me to write this book. In addition, I wish to thank Alice Oven and Tasha D'Cruz of Imperial College Press for their highly professional service and help in publishing the book. I also wish to thank my colleagues at Southampton: Keith Jones, Jacky Lumby and Daniel Muijs for their helpful suggestions for this book.

Finally, I would like to express my gratitude to the University of Chicago School Mathematics Project Royalty Fund and the Academic Research Fund of the National Institute of Education, Nanyang Technological University for their sponsorship in grants for me to conduct the Chicago and Singapore studies, and to the University of Southampton and its Southampton Education School for providing a supportive academic environment for me to complete this book. Of course, I must say that I am entirely responsible for all the views expressed in the book.

Lianghuo Fan
University of Southampton
Southampton Education School
March 2014

Part I

The Chicago Study

Chapter 1

Introduction

Background of the Chicago Study

Since the early 1980s, when the US started another wave of educational reform and improvement from the momentum of the most notable report, *A Nation At Risk: The Imperative for Educational Reform* (National Commission on Excellence in Education, 1983), dozens of influential reports, like the Holmes Group's trilogy (1986, 1990, & 1995), have stressed the importance of teachers and teachers' knowledge on students' learning. In practice, however, the issues of teachers did not receive significant attention from policy makers and educational reformers during the school reform movement in a long time. For example, in the well-publicized Chicago public school reform initiated in the late 1980s, while the focus was on improving school governance, and the dominant reform strategy was decentralization (and later integrated governance), how to improve teachers' teaching quality in the classrooms was largely neglected in the reform initiatives (e.g., see McKersie, 1996; Wong *et al.*, 1997). As a matter of fact, in the report of *What Matters Most: Teaching for America's Future*, the National Commission on Teaching and America's Future (1996) concluded,

> On the whole, the school reform movement has ignored the obvious: What teachers know and can do makes the crucial difference in what children learn. (p. 5)

Why did the school reformers fail to address the issues of teachers? Researchers have suggested that one main reason, among others, is that solid knowledge about teacher improvement is lacking; people still do not know much about the nature of teaching as an occupation and how to effectively improve the teaching taskforce (e.g., see Dreeben, 1996). In the case of mathematics, Jack Price, the then president of the National Council of Teachers of Mathematics (NCTM), pointed out,

> We still have not developed a set of effective models that will help us prepare, and become, better teachers. This is high on the list of needs to be addressed. (Price, 1996, p. 3)

Recognizing the overall need for seeking knowledge of teacher improvement and focusing on teachers of mathematics, the study presented in the first part of this book particularly aims to examine, in the domain of pedagogy, "How do teachers develop their knowledge?" As the study was conducted in the metropolitan area of Chicago, we shall hereafter call it "the Chicago study", for short.

Need for the Study

In terms of research, as more and more studies have been done in the area of teacher knowledge since the early 1980s (see more in Chapter 2), a number of researchers have begun, from different perspectives, to recognize the need for addressing the issue of how teachers develop their knowledge.

After finding that there existed considerable differences in teachers' conceptions of mathematics and mathematics teaching, Thompson (1984) claimed,

> As more is learned about teachers' conceptions of mathematics and mathematics teaching, it becomes important to understand how those conceptions are formed and modified. Only then will the findings be of use to those involved in the professional preparation of teachers, attempting to improve the quality of mathematics education in the classroom. (p. 127)

Shulman (1986a) raised the same question with a broader meaning in his presidential address at the 1985 annual meeting of the American Educational Research Association,

> What are the sources of teacher knowledge? ... when did he or she come to know it? How is new knowledge acquired, old knowledge retrieved, and both combined to form a new knowledge base? (p. 8)

In a relatively comprehensive review of the research on teachers' knowledge and its impact, Fennema and Franke (1992) pointed out,

> [T]he challenge is to understand knowledge as it grows and changes and to discover what experiences [of teachers] contribute to this growth and change. (p. 161)

In another comprehensive review of research and teacher education, Cooney (1994) stated,

> One could argue that the means by which teachers learn such (pedagogical content) knowledge is one, if not the, defining point for teacher education and consequently should be the focal point of research on teacher education. (p. 611)

Other researchers have listed how teachers develop, or construct, their knowledge as one of the central questions for teacher educators (Johnston, 1992, p. 123) and policy makers (Grossman, 1994, p. 6121).

The Chicago study is in part motivated by sharing those scholars' thoughts. Through investigating two specific research questions, it is hoped that the study can obtain a relatively systematic picture of how mathematics teachers develop their pedagogical knowledge, and therefore explore its implication for teachers' professional development and shed light on how to effectively improve teachers' pedagogical knowledge.

Statement of the Problem

The study focuses on teachers' knowledge in the domain of pedagogy. The general research question of this study is,

> How do teachers develop their pedagogical knowledge?

More specifically, the study is intended to address the following two questions:

(1) Are there different sources of teachers' pedagogical knowledge?

Although the answer to this question seems obvious, people have traditionally assumed with little empirical study that "teachers teach the way they were taught" and emphasized the influence of teachers' previous experience as students on their knowledge of pedagogy (e.g., see Lortie, 1975, Chapter 3; NCTM, 1989a, p. 62, 1991, p. 124). I doubt that assumption and hope this study can empirically identify what are the major sources for teachers to develop their pedagogical knowledge.

(2) How do different sources contribute to the development of teachers' pedagogical knowledge?

In other words, if there are different sources from which teachers learn how to teach mathematics, what is the relative importance of these sources to the development of their pedagogical knowledge?

Structure of Part I

The first part of the book contains nine chapters, presenting the Chicago study in a fairly detailed and comprehensive way.

Chapter 2 first introduces some epistemological background for the study about the concept of knowledge, and then provides a relatively broad review of the relevant literature in the field of teacher knowledge, which is divided around three major issues: what knowledge do teachers need, what knowledge do teachers have, and how do teachers develop their knowledge.

Chapters 3 and 4 present the research background necessary to interpret and understand the findings of the study. Drawing partially on the previous research reviewed in Chapter 2, Chapter 3 establishes a conceptual framework of teachers' pedagogical knowledge and the possible sources for them to develop such knowledge. Chapter 4 describes the research design and the procedures used in the study, including the development of instruments, which was largely based on the conceptual framework presented in the previous chapter.

The following three chapters, Chapters 5, 6, and 7, report the core findings of the study. These chapters report the results and analyses about how teachers developed their pedagogical curricular knowledge, pedagogical content knowledge, and pedagogical instructional knowledge, respectively.

Chapter 8 supplements the three previous chapters by looking at some other issues, which are not included in the previous chapters, but are related and helpful for us to understand the issues addressed in those chapters from a different angle. The chapter focuses on specific sources and does not separate the general pedagogical knowledge into three different components.

Chapter 9 provides a summary of the findings and the conclusions obtained from the Chicago study. It also discusses the implications of the findings for teacher educators, school administrators, and teachers themselves on how to effectively pursue the improvement of teachers' pedagogical knowledge. Recommendations for further research are also made at the end of the chapter.

Chapter 2

Review of the Literature

Since the early 1980s, researchers in various education-related disciplines including teacher education, educational policy, psychology, and curriculum and instruction of different subjects such as English, mathematics, science, and physical education, have showed rapidly increasing interest in the field of teacher knowledge.[1] However, the attention paid to the field has been unbalanced, and many questions have still not been well answered or even touched upon.

Generally speaking, there exist mainly three major issues in the field of teacher knowledge: 1. What knowledge do teachers need? 2. What knowledge do teachers have? 3. How do teachers develop their knowledge?[2]

While the Chicago study focuses on how teachers develop their knowledge in the domain of pedagogy, the review herein also includes the

---

[1] Among many other researchers, there were three researcher groups: Shulman and his colleagues and a group of students based at Stanford University with the main focus on the growth of beginning teachers' knowledge; Connelly and Clandinin and their colleagues and students mainly based at the University of Toronto on teachers' personal practical knowledge; and Feiman-Nemser and Ball and many others based at Michigan State University on teacher learning, whose large amount of work and publications on teacher knowledge were particularly noticeable (e.g., see Shulman, 1987; Ball, 1989; Feiman-Nemser & Parker, 1990; Lappan & Theule-Lubienski, 1994; Clandinin & Connelly, 1995).

[2] There are also studies on "how do teachers use their knowledge in practice, or what are the influences of teachers' knowledge?" and studies on "how to understand teachers' knowledge". In some sense, the former can be eventually related and reduced to the first issue "what knowledge do teachers need", and the latter to the second issue "what knowledge do teachers have". However, the three issues listed here cannot be logically reduced to one another.

literature on the first two issues. In doing so, I hope to provide a relatively broad background for this study and explain where the study fits into the body of previous work and how the study will contribute to it. In addition, the literature on the first issue has been important for me to establish this study's conceptual framework, which will be described in Chapter 3.

Before getting to the studies of teacher knowledge, let me briefly introduce some epistemological background about the concept of knowledge, which I think has been largely neglected in a considerable number of studies on teacher knowledge and caused some substantial problems and difficulties in advancing the research in this area.

What is Knowledge?

Historically, the nature of knowledge has received a great deal of attention from philosophers since the ancient Greeks. Many great thinkers, such as Plato, Aristotle, Descartes, Kant, Russell, and Dewey, to name a few, have made important contributions to the theory of knowledge (e.g., see Chisholm, 1966; Quinton, 1967; Potter, 1987). In recent times, epistemologists and other scholars are still studying and producing knowledge about knowledge.

Although "knowledge" is a commonly used word, and it seems that anyone who uses this word roughly knows what it means, philosophers or epistemologists have shown us that it is very difficult, if not impossible, to give a precise definition of knowledge. Russell called the word "knowledge" a "highly ambiguous" one in his 1913 manuscript (Russell, 1992, Chapters IV & VI), and later explicated again that it is a term "incapable of precision" (Russell, 1948, p. 516). And, throughout *Knowing and the Known*, Dewey and Bentley called the word a "loose name" or a "vague word" (Dewey & Bentley, 1949). Other scholars such as Wilson (1926) and Prichard (1950) held that knowledge is a primitive and indefinable concept (also see Laird, 1930, pp. 113–121).

Despite the difficulty in reaching a universal definition, people have developed and employed various definitions, or (it is perhaps more appropriate to say) analyses, of "knowledge" from different perspectives to understand what knowledge is.

In epistemology, the term "knowledge" is traditionally most widely defined as "justified true belief", in contradistinction with "error", "false belief", or "mere personal opinion" (such as "guess") (e.g., see Quinton, 1967; Shope, 1983; Fenstermacher, 1994). That is, if one claims he/she knows that is the case, there must exist three preconditions: 1. he/she must believe that is the case; 2. indeed, that is the case; and 3. he/she can justify that is the case.

For example, within mathematical knowledge, if one is to be said to know that "$x - y$ is a factor of $x^3 - y^3$", he/she must believe that "$x - y$ is a factor of $x^3 - y^3$", it must be true that "$x - y$ is a factor of $x^3 - y^3$", and he/she must have justification, e.g., being able to prove that $(x^3 - y^3) = (x - y)(x^2 - xy + y^2)$ in believing that "$x - y$ is a factor of $x^3 - y^3$".

This standard analysis of "knowledge" was especially popular among great philosophers before the 19th century (e.g., see Laird, 1930, p. 93). After that, a trend seems to have emerged among epistemologists that enlarged the definition of knowledge from "justified true belief" to "objectively grounded, or adequately supported, belief", or in other words, "objectively reasonable belief", or "evidentially supported belief" (e.g., see Wolf, 1921; Green, 1971; Jackson, 1986, p. 91; Fenstermacher, 1994). According to this "softer" standard, a claim made by someone can be counted as his knowledge if he has reasonable (or acceptable) evidence, either directly or indirectly (Chisholm, 1966), to establish the claim. For example, a teacher can claim to know that group discussion is an effective way for classroom teaching, if he has reasonable evidence, such as that he tried several other teaching methods and found that students paid more attention to their learning when using group discussion.

In a certain sense, the above definitions of "knowledge" only apply to propositional knowledge, that is, believing that "something is the case". However, more broadly, especially in ordinary discourse, the term "knowledge" is used in a variety of senses in different contexts. People may claim they know an object (e.g., a person, a place, or a thing) in the sense of "being acquainted with it" or "being familiar with it"; for example, someone claims that he knows Chicago because he lived in Chicago for several years and has been familiar with the city. Again, people may claim that they know something (e.g., a story, a book, or an event) in the sense of "being aware of it" or "having heard of it"; for example, one claims he knows that World

War II ended in 1945 because he heard of it from his teachers. Moreover, people may claim that they know how to do an activity (e.g., drive a vehicle, cook, or conduct a medical operation) in the sense of "remembering how" and/or "having acquired relevant skills or ability" to conduct the activity (e.g., see Potter, 1987).

The above explanations of "knowledge" imply that there are different kinds of knowledge. It seems to me that being aware there are different kinds of knowledge is helpful, and often necessary, to keep a balanced view of "what is knowledge".[3] As Pears pointed out:

> There are so many varieties of knowledge, and each of them has so many aspects that it is easy to neglect some of the phenomena and to produce a theory that only covers part of the field. One way of making sure that nothing is neglected is to begin with a rough classification of different types of knowledge. (1971, p. 4)

Epistemologically, Russell once distinguished two kinds of knowledge: *knowledge of truths* (roughly equivalent to propositional knowledge) and *knowledge of things* (1959). Furthermore, for "knowledge of truths", he discerned a distinction between *immediate* (or *direct, intuitive*) *knowledge*, which is conveyed in intuitive statements such as direct judgment of perception and mathematical axioms; and *mediate* (or *indirect, inferential, derivative*) *knowledge*, which is conveyed in demonstrable necessary propositions and inferred empirical statements such as mathematical theorems. For "knowledge of things", he further drew a distinction between *knowledge by acquaintance* and *knowledge by description*. According to him, knowledge by acquaintance means being directly aware without the intermediary of any process of inference or any knowledge of truths. For instance, one will immediately have knowledge of a table's color and shape when he/she sees that table. However, his/her knowledge of the table itself as a physical object is not direct knowledge; it is obtained through acquaintance with

---

[3] Fenstermacher (1994) showed in his review from the perspective of epistemology that some literature on teacher knowledge lacks such a balanced view of knowledge, and hence carries "serious epistemological problems". However, he included three kinds of literature: on knowledge by teacher, on knowledge about teacher, and on knowledge about teaching, under the same umbrella of "teacher knowledge", which seems to me also problematic (see more in the section "Teachers' Knowledge" in Chapter 3).

the sense-data that make up the appearance of and are used to describe the table. It is *knowledge by description*.

Ryle (1949) made another important distinction between *knowing how* and *knowing that*. According to Ryle, *knowing that* is equivalent to propositional knowledge, or knowledge of truths, mentioned above. He argued that traditional theorists, particularly the ancient Greeks, had been too much preoccupied with human capacity to obtain knowledge of truths, ignoring and despising the capacity to perform tasks, which *knowing how* is about.[4] Stressing the importance of human knowledge as *knowing how*, Ryle maintained,

> In ordinary life, on the contrary, as well as in the special business of teaching, we are much more concerned with people's competencies than with their cognitive repertoires, with the operations than with the truth they learn. Indeed even when we are concerned with their intellectual excellencies and deficiencies, we are interested less in the stocks of truths that they acquire and retain than in their capacities to find out truths for themselves and their ability to organise and exploit them, when discovered. (p. 28)

Although some scholars like Johnson (1989) have questioned Ryle's dichotomy, the dichotomy seems to me important to be noted when we are concerned with either teachers' knowledge, as in this study, or learners' knowledge, as mentioned by Ryle.

Other distinctions made by scholars from various perspectives and for different purposes seem, to me, less fundamental and often less strict, yet often easier to understand and more practical. For example, people sometimes distinguish theoretical knowledge from practical (or empirical) knowledge,[5] craft knowledge from research knowledge, conceptual knowledge from methodological (or procedural, or process) knowledge,

---

[4]According to Doren, before Thales, most human knowledge had been *knowing how*, comprising pragmatic rules for success in enterprises from hunting to growing crops, from organizing households to governing cities, from creating art to waging war. It is the Greeks who first began to philosophize about the nature of things. For a more detailed discussion about the role the ancient Greeks played in advancing human knowledge, see *A History of Knowledge* (Doren, 1991, Chapter 2).

[5]It has been not uncommon that different researchers have employed different terms for the same kinds of knowledge, or employed the same terms for different kinds of knowledge. See more below in this section.

professional knowledge from general knowledge,[6] and even personal or local knowledge from public knowledge (e.g., see Machlup, 1980; Love, 1985; Gilbert, Hirst, & Clary, 1987; Leinhardt, 1990; Grimmett & Mackinnon, 1992; Cochran-Smith & Lytle, 1993, Chapter 3; Eraut, 1994).

It is important to note that when we talk about knowledge there are three components involved: the knower — the subject of knowledge (who knows); the known — the object of knowledge (what is known); and the knowing — the interaction of the subject and object (how to know).

Therefore, we can roughly see that the majority of the distinctions mentioned above are based on the known, such as *knowing that* and *knowing how*, *theoretical knowledge* and *practical knowledge*, *self-knowledge* and *external world knowledge*. Some are based on the knowing, such as *knowledge by description* and *knowledge by acquaintance*, *craft knowledge* and *research knowledge*, *direct knowledge* and *indirect knowledge*. Still some are based on the knower, such as *professional knowledge* and *general knowledge*, *personal knowledge* and *public knowledge*. Obviously, more distinctions and sub-distinctions are possible. I will mention some other types of knowledge later in this study.

In my view, a subject's knowledge of an object is a mental result of the interaction of the knower and the known, or the knowing. Further distinction is possible based on the forms of the mental result. For example, Polanyi (1966) made another important distinction between *tacit knowledge*, referring to the knowledge people have in their minds but cannot tell in words, and the other kind of knowledge, which some researchers termed as " *explicit knowledge*" (Alexander, Schallert, & Hare, 1991).

In the educational community, many studies on knowledge, including on teachers' knowledge, in the last decades simply failed to explicitly explain what is meant by the term "knowledge". Implicitly, as Alexander, Schallert, and Hare revealed in a review: "it goes nearly without saying that knowledge refers to an individual's personal stock of information, skills, experiences, beliefs, and memories" (1991, p. 317). Although their

---

[6]This is a distinction based on the knower (professionals or general public). Sometimes, people use "general knowledge" to contrast with "situated knowledge" (e.g., Orton, 1993), which is based on the known. The meaning of the knower and the known is discussed in the following paragraph.

review focused on research in learning and literacy, it seems to me that their conclusion can be largely applied to the research on teacher knowledge. Because of lacking adequate epistemological consideration, it is not strange that quite a few studies even mixed up teachers' knowledge with their beliefs, or skills, or their personal experiences without making necessary explanation about the relations and differences between those concepts (e.g., see Clandinin & Connelly, 1987; Fenstermacher, 1994; Fang, 1996). Wideen, Mayer-Smith, and Moon contended that some shortcomings of the research on teacher (professional) knowledge "arise when one takes an uncritical and unproblematic view of knowledge" (1996, p. 187).

For those studies that do explicitly explain what knowledge means for them, the definitions of knowledge are often more utility oriented, rather than epistemologically oriented. For example, in Alexander *et al.*'s review, "knowledge encompasses all that a person knows or believes to be true, whether or not it is verified as true in some sort of objective or external way" (1991, p. 317). Hood and Cates (1978) stated that "knowledge refers to the total body of data, information, intelligence, and technology, and their organizing structures and principles (i.e., the sum of all that is known)" (p. 4; cited in Love, 1985).

Regarding the types of knowledge, Alexander *et al.* listed more than 20 types of knowledge (1991, pp. 332–333) employed by educational researchers, and concluded that there were four main problems across the studies they reviewed,

> (a) subcategories of knowledge were inconsistently incorporated;[7] (b) different aspects of knowledge were referred to by the same terms; (c) the same aspects of knowledge were referred to by different terms, and (d) the interactions among the different aspects of kinds of knowledge were represented differently, or ignored altogether. (p. 319)

It seems to me that (a) and (d) are more fundamental problems, though the biggest problem I think is that many researchers (including those on teacher knowledge) rarely took the well-established epistemological distinctions into account when classifying different kinds of knowledge.

---

[7]Unfortunately, they also did not consistently use their own definition of knowledge as beliefs (see the previous paragraph) in their review. For instance, they also included *knowing how* (such as procedural knowledge), which is not as the same as beliefs.

In this study, I basically define "knowledge" as a mental result of certain interaction of the knower and the known; knowledge can be of various forms and types, including *knowledge of truths* and *knowledge of things*, *knowing that* and *knowing how*, *tacit knowledge* and *explicit knowledge*, and *direct knowledge* and *indirect knowledge*. In short, under this definition, teachers' pedagogical knowledge include all teachers know about the pedagogy for their classroom teaching; with teachers being the knower, pedagogy for their teaching being the known, and interaction being a process of teachers producing their knowledge or a process of learning or accepting others' knowledge about teaching (see more details in Chapter 3).

Now let us turn to a review of recent studies on teacher knowledge. By "recent studies" I mainly focus on, especially for the last two sections of this chapter, empirical studies since 1980, although there are also theoretical and analytical studies available, and there existed a number of studies in this line before 1980, which will also receive some attention in this study when appropriate.

What Knowledge do Teachers Need?

As Znaniecki said that "every individual who performs any social role is supposed ... to possess the knowledge indispensable for its normal performance" (1965, p. 24), it is obvious that everyone will agree that teachers must need certain knowledge in order to teach effectively. However, there has been no consensus on "what kind" and "what amount" of knowledge teachers need, or in other words, is indispensable for teachers' normal performance.

With respect to "what kind" of knowledge teachers need, people have traditionally thought what teachers needed to know was basically what they would teach. This belief has been around since the ancient times. In other words, teachers need subject matter knowledge. According to Monroe's *A Cyclopedia of Education*, this belief was practically held until the 18th century, several centuries after "pedagogy" began to receive more and more attention following the Renaissance. From the early 19th century, "it came to be an accepted principle that (elementary) teachers should know not only the subjects they were to teach, but also the art of their craft" (Monroe, 1913, p. 622), that is, in addition to subject matter knowledge, teachers also need pedagogical knowledge. Nonetheless, on the whole, teacher knowledge, as well as teacher education, did not receive much attention as a field of intellectual inquiry prior to the 1980s (Carter, 1990; Cooney, 1994).

Teacher knowledge became a focal point in the rapid increase of research on teacher education following the early 1980s. Researchers have considerably expanded the conception of teacher knowledge, and utilized various terms, most commonly, "knowledge base for teaching", "professional knowledge", and "practical knowledge", and generated different models to explore what knowledge teachers need for their effective teaching.

According to Elbaz (1981, 1983), teachers need a broad range of knowledge, which she termed as "practical knowledge", when confronting "all manner of tasks and problems", to guide their work: knowledge of subject matter; knowledge of curriculum (about the structuring of learning experiences and curriculum content); knowledge of instruction (about classroom management, instructional routines, students' needs, abilities, and interests); knowledge of the milieu of schooling (about the social framework of the school and its surrounding community); and knowledge of self (about their own strengths and shortcomings as teachers).[8]

Shulman (1987) defined seven categories of knowledge which constitute the knowledge base of teaching: (1) content knowledge, which mainly refers to subject matter knowledge;[9] (2) general pedagogical knowledge, with special reference to those broad principles and strategies of

---

[8]Elbaz used "practical knowledge" because "this term appears to open up a wide range of possibilities in looking at teachers. It clearly reflects the fact that teachers' knowledge is broadly based on their experiences in classrooms and schools and is directed toward the handling of problems that arise in their work.... [T]he term reminds us that what teachers know is capable of being formulated as 'knowledge',... and being used to generate consistent practice" (1981, p. 67). It seems to me this conception is problematic; also see Fenstermacher (1994).

[9]Shulman changed and modified his categorization and definition several times. According to a previous explanation (Shulman, 1986b), there are three categories of content knowledge: (a) subject matter knowledge referring to the comprehension of the subject appropriate to a content specialist in the domain; (b) pedagogical knowledge — referring to the understanding of how particular topics, principles, strategies, and the like in specific subject areas are comprehended or typically misconstrued, are learned and likely to be forgotten; and (c) curricular knowledge — referring to the familiarity with the ways knowledge is organized and packaged for instruction, in texts, programs, media, workbooks, other forms

classroom management and organization that appear to transcend subject matter; (3) curriculum knowledge, with particular grasp of the materials and programs that serve as "tools of the trade" for teachers; (4) pedagogical content knowledge, which represents the blending of content and pedagogy into an understanding of how particular topics, problems, or issues are organized, represented, and adapted to the diverse interests and abilities of learners, and presented for instruction; (5) knowledge of learners and their characteristics; (6) knowledge of educational contexts, ranging from the workings of the group or classroom, the governance and financing of school districts, to the character of communities and cultures; and (7) knowledge of educational ends, purposes, and values, and their philosophical and historical grounds.

Noticeably, there have been a considerable number of studies focusing on pedagogical content knowledge of different subject matters since Shulman's presidential address to the American Educational Research Association in 1985 where he drew widespread attention to this kind of knowledge, though some researchers have modified his definition for different purposes (e.g., Grossman, 1988; Marks, 1990; Cochran, DeRuiter, & King, 1993; Even, 1993; Meredith, 1993; Neagoy, 1995; Griffin, 1996; Carpenter, Fennema, & Franke, 1997). In fact, the theme of one issue of the *Journal of Teacher Education* (Vol. 41, No. 3) was devoted to "pedagogical content knowledge".

Gilbert, Hirst, and Clary (1987) proposed a broad taxonomy, both sequential and hierarchical, of a professional knowledge base for "classroom teachers". Admitting not taking into account the knowledge of content in Shulman's list, Gilbert *et al.* designed four levels in their taxonomy. The first level is knowledge of school as an institution, including knowledge of history of American education, philosophy of education, professional ethics, public policy, school law, and school organization. The second level is knowledge of students, consisting of knowledge of multicultural education, socioeconomic factors, learning theory, and human development. The third level is knowledge of teaching, including knowledge of curriculum development, teaching methods, educational technology, measurement, and

---

of practice, and the like. However, he later employed the term "pedagogical content knowledge" to replace "pedagogical knowledge" in (b) (Shulman, 1986b). And, here (b) and (c) are listed independently, not a subset of "content knowledge".

learning styles. The last is the decision-making level, called knowledge of clinical applications, under which are listed knowledge of human relations, educational management, evaluation, and modeling.

Drawing on Shulman's definition, Grossman (1988, 1991) divided teacher knowledge into four types: "general pedagogical knowledge", "subject matter knowledge", "pedagogical content knowledge", and "knowledge of context". While having more or less employed Shulman's definition for the other three types of the knowledge, she particularly referred "subject matter knowledge" to not only knowledge of the content of a subject area, but knowledge of the substantive and syntactic structures of the discipline;[10] and later, she also included "beliefs about subject matter" as a component of subject matter knowledge (Grossman, Wilson, & Shulman, 1989).

Cochran, DeRuiter, and King (1993) proposed an integrative model of "pedagogical content knowing" which originated from and refined Shulman's concept of pedagogical content knowledge for teacher preparation. According to them, there are four components of teacher knowledge which comprise pedagogical content knowing: knowledge of subject matter; knowledge of pedagogy; knowledge of students; and knowledge of environmental contexts. They emphasized knowing and understanding as active processes and the simultaneous development of all aspects of knowing how to teach in teacher preparation.

In the case of mathematics as a subject, Leinhardt and Smith (1985) argued that there were two core areas of teacher knowledge: lesson structure knowledge, and subject matter knowledge. The former includes the skills needed to plan and run a lesson smoothly, to pass easily from one segment to another, and to explain material clearly. The latter includes mathematics concepts, procedures, and their connections, the understanding of classes of students' errors, and curriculum presentation. They further explained that subject matter knowledge acts "as a resource in the selection of examples, formulation of explanations, and demonstration" (p. 247). Their definition of lesson structure knowledge is similar to that of "general pedagogical

---

[10]Grossman used Schwab's definitions of "substantive" and "syntactical" knowledge. While substantive knowledge largely refers to the body of interrelated concepts of a discipline, syntactical knowledge refers to the methods used in the discipline to construct knowledge (Schwab, 1964, 1968).

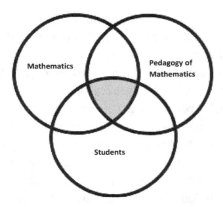

Figure 2.1 Three types of knowledge for mathematics teachers by Lappan and Theule-Lubienski (1994).

knowledge" defined by others such as Shulman, and their definition of "subject matter knowledge" contains both of what Shulman termed "content knowledge" and "pedagogical content knowledge".

In the Seventh International Congress on Mathematical Education held in 1992, Lappan and Theule-Lubienski (1994) emphasized that teachers needed at least three kinds of knowledge in order to be effective in choosing worthwhile tasks, orchestrating discourse, creating a learning environment, and analyzing teaching and learning. They represented those kinds of knowledge in a Venn diagram (Figure 2.1)

They argued that "teachers work in the intersection of these domains of knowledge. It is the interplay of various considerations that leads to defensible pedagogical reasoning on the parts of teachers" (Lappan & Theule-Lubienski, 1994, p. 253).

Fennema and Franke (1992) also developed a model for examination and discussion centered on teachers' knowledge as it occurs in the context of the classroom. The model, which emphasizes the interactive and dynamic nature of teachers' knowledge, has four components: knowledge of mathematics; (general) pedagogical knowledge; knowledge of learners' cognition in mathematics; and context-specific knowledge, which indicates the teachers' knowledge in context or as situated.

A more comprehensive categorization was done by Bromme (1994). He distinguished five fields of teachers' professional knowledge that are needed for teaching (p. 85): (a) knowledge about mathematics as a discipline, which contains mathematical propositions, rules, mathematical modes of thinking, and methods; (b) knowledge about school mathematics, which has a "life of its own" with its own logic, not just simplifications of mathematics as it is taught in universities; (c) the philosophy of school mathematics, which are ideas about the epistemological foundations of mathematics and mathematics learning and about the relationship between mathematics and other fields of human life and knowledge; (d) general pedagogical (and psychological) knowledge, which mainly includes general classroom organization and instructional communication, and has a relatively independent validity separate from the school subjects; and (e) subject-matter-specific pedagogical knowledge, similar to what Shulman termed "pedagogical content knowledge".

I would like to point out several things here. First, it is surprising that, except Gilbert *et al.*, few of the researchers above have explicitly recognized the importance of teachers' knowledge of technology to their teaching. (By technology I mean "technological equipment", most commonly and importantly including calculators, computers, and software. See more in Chapter 5.) As technology has deeply affected both what to teach and how to teach, it seems necessary to include technological knowledge and locate it appropriately in the structure of teachers' knowledge.

Second, in general, the conceptual models of what kind of knowledge teachers need described above are largely ideological and "artificial" (Gilbert *et al.*, 1987) rather than empirical; they largely reflected researchers' own beliefs of education,[11] experience, expertise, and research interests and fields. This is the basic reason that there is so much difference in those models. In some sense, it is not difficult to make a general list of what knowledge teachers should have, but without empirical evidence of

---

[11] For example, if one's educational goal of teaching a subject is understanding such as defined by NCTM's *Curriculum and Evaluation Standards for School Mathematics* (1989b), he/she will understandably emphasize that teachers first need a deep understanding of the subject. In fact, the NCTM *Standards* have influenced people's conception of what mathematics teachers need to know in order to meet the challenge (e.g., see Lappan & Theule-Lubienski, 1994; Talbert, McLaughlin, & Rowan, 1993). Another example that Ladson-Billings emphasized, from the perspective of Black feminist thought, is that teachers needed to have knowledge of "culturally relevant pedagogy" (1995).

its importance to teaching, it is not so easy and so convincing to answer why teachers need this kind of knowledge. From that point of view, it is necessary to somehow adopt a more empirically based approach to the issue in further studies, though it may be much more methodologically difficult.[12]

Last, since researchers have had so many different conceptualizations and even conflicting constructs about teachers' knowledge, more exchanges between researchers are needed in order to search common ground and establish some kind of consensus in this issue, as with other issues in the research of teacher education (Cooney, 1994; Grossman, 1994).

Despite the difference in researchers' models of what knowledge teachers need, there exist two common, though sometimes termed differently, essential kinds of knowledge, that is, "subject matter knowledge", and "pedagogical knowledge". This fact largely suggests the central importance of teachers' subject matter knowledge and pedagogical knowledge to their teaching. It is partly because of the importance of pedagogical knowledge that this study focuses on teachers' knowledge in this domain.

The issue of "what amount" of knowledge is needed by teachers is more difficult to address and, perhaps because of its difficulty, has been much less addressed. There is a Chinese proverb, "if you want to give students a cup of water, you need to have a bucket of water". It clearly suggests that teachers need to know more than students in the same kind of knowledge ("water"). However, it is unclear how much more is really enough. As a matter of fact, focusing on subject matter knowledge, Begle (1979) pointed out,

> It seems to be taken for granted that it is important for a teacher to have a thorough understanding of the subject matter being taught. The question is never raised . . . as to how thorough the understanding needs to be. (p. 28)

In addition, Leinhardt and Smith (1985) argued,

> It is surprising that so little research has explored the type and level of subject matter skill used and required by teachers. With few exceptions, research in

---

[12] Since the 1980s, there have been numerous empirical studies, usually on a small scale, on the impact of teachers' particular knowledge on their classroom teaching and student learning, which is directly or indirectly related to the question, "Why do (or do not) teachers need this kind of particular knowledge?" (e.g., Carlsen, 1988; Brickhouse, 1990; Gudmundsdottir, 1990). However, more comprehensive and systematic studies are still needed.

teachers' subject matter knowledge (level, organization, and understanding) has been alluded to but not studied. (p. 248)

Theoretically, for any kind of knowledge discussed above, we can always ask, in one way or other, "what amount of this kind of knowledge (e.g., pedagogical knowledge) do teachers need?"[13] However, traditionally, researchers focused their attention in this issue on teachers' general knowledge, and more often, only on subject matter knowledge, even though the study in this area was still relatively scant, as cited above.

With a teacher's knowledge quantitatively measured in terms of his/her educational attainment such as the total number of relevant courses he/she took, or his/her grade point average, or even his/her score on a standardized achievement test, most studies have consistently showed that there was statistically little relationship between amount of teachers' knowledge and teaching effectiveness in terms of students' scores on standardized tests (e.g., see Begle, 1972, 1979; Copeland & Doyle, 1973; Eisenberg, 1977). This consistent result led Begle (1972) to suggest that teachers may only need a certain amount of subject matter knowledge; beyond a reasonable threshold, further subject matter expertise does not much matter. Later, Begle (1979) even strongly discouraged further studies in this direction after a comprehensive review of relevant studies, as he concluded:

> Probably the most important generalization which can be drawn from this body of information is that many of our common beliefs about teachers are false, or at the very best rest on shaky foundations.... [T]he effects of a teacher's subject matter knowledge and attitudes on student learning seem to be far less powerful than most of us had realized.... These numerous studies have provided us no promising leads.... Our efforts should be pointed in other directions. (pp. 54–55)

Notably, there is a similar study done by Mullens, Murnane, and Willett (1996) in Belize. Using a statistical model relating 1,043 third-grade students' mathematics achievement and their 72 teachers' characteristics, they found that teacher knowledge of mathematics (measured on a

---

[13] Some researchers have questioned if knowledge, particularly non-propositional knowledge, consists of measurable units (e.g., Eisner, 1997). To me, however, the question is not if knowledge is measurable, but how knowledge can be measured. If we admit there is a difference between different people's knowledge, then there is a way to detect the difference, quantitatively or qualitatively.

primary-school-leaving exam) and educational certification (whether or not completed high school) were significantly related to student learning,[14] but teacher training (whether or not completed a three-year training program) was not. For this insignificance of the teacher training, they admitted that a more appropriate methodology might be needed.

Generally speaking, today this kind of study has been thought by many researchers to be not only of little use, but naïve (e.g., Even, 1993; Cooney, 1994). Researchers have pointed out that teachers' subject matter knowledge and students' achievement may have been inadequately conceptualized in those studies. The subject knowledge teachers actually have cannot totally be reflected in the number of courses taken in college, or grade point average, or scores on a standardized test; students' achievement is also more than what is measured in a standardized test, and their gain-scores can only partially represent teachers' teaching effectiveness; additionally, there are many other contexts and conditions that temper any possible relationship between teachers' knowledge and students' learning (Grossman, Wilson, & Shulman, 1989; Even, 1993; Cooney, 1994).

Since the 1980s, researchers have used more qualitative approaches to analyze teachers' knowledge and its influence on classroom teaching, and argued that teachers' subject matter knowledge of correct facts, concepts, theories, and processes is not enough. Beyond that, teachers also need to know the nature, the structure, and the epistemology of the discipline, and its existence in culture and society (Thompson, 1984; Leinhardt & Smith, 1985; McDiarmid, 1988; Ball, 1990a, 1991; Even, 1993; Tate, 1994; Baturo & Nason, 1996).

The amount of general knowledge teachers have was at one time represented by their educational level, namely, the degrees they hold. Overall, although there were exceptions, the results of most studies showed there was no, or a modest, relationship between higher degrees teachers held and the achievement their students obtained (e.g., see Bidwell & Kasarda, 1975; Turner *et al.*, 1986; Turner, 1990), similar to what we mentioned above regarding teachers' subject matter knowledge. Also, there exist the similar methodological problems.

---

[14] Note one reason for the existence of the strong relationship may be that those teachers' knowledge and certification are within the so-called "threshold" by Begle (1972).

The issue of what amount of knowledge teachers need has an important implication for educational policy. If Begle's suggestion that beyond a reasonable threshold further subject matter expertise does not much matter is correct, then from a policy perspective it is largely not worthwhile for a teacher to pursue an educational degree higher than a certain level, if the economic expenses are relatively high. Actually, this is a main reason that Knapp *et al.* (1990) argued that "a universal Master's degree requirement is an inefficient method for improving elementary and secondary education (in Virginia)." Though, on the other hand, from the perspective of historical trends and social expectation, requiring teachers to have higher degrees seems unavoidable (Turner, 1990).

What Knowledge do Teachers Have?

Because there have been many studies investigating what knowledge teachers of different school subjects actually have, this section mainly focuses on mathematics. By and large, studies have concentrated on two kinds of teachers' knowledge: subject matter knowledge and pedagogical knowledge, of which pedagogical content knowledge has received much attention since the mid-1980s.

Most studies investigating teachers' knowledge have been limited to some specific domains of a subject matter, such as, in mathematics, arithmetic operations, function, and limit. In the domain of fractions, Leinhardt and Smith (1985) conducted an in-depth study examining the subject matter knowledge, which partly included pedagogical content knowledge as I described earlier, possessed by eight fourth-grade mathematics teachers: four expert teachers and four novice teachers. They found that although the four experts generally had a better knowledge structure than novices, there were considerable differences among those experts: two experts had relatively high-level mathematics knowledge about fractions, one had middle-level knowledge, and one had barely sufficient mathematics knowledge for classroom instruction. In addition, there were differences among the experts' "pedagogical content knowledge" in the level of conceptual information presented as well as in the degree to which procedural algorithmic information was presented.

Carpenter et al. (1988) studied 40 first-grade teachers' pedagogical content knowledge of children's solutions of addition and subtraction word problems, and found that most teachers could identify many of the critical distinctions between problems and the primary strategies used by children to solve different kinds of problems, but this knowledge generally was not organized into a coherent body that related distinctions between problems, children's solutions, and problem difficulty.

Ball (1990b) investigated 19 prospective elementary and secondary teachers' understanding of division. She concluded that, although many of them could produce correct answers, several could not, and few were able to give mathematical explanations for the underlying principles and meanings. Those prospective teachers' knowledge was generally fragmented, and each case of division was held as a separate bit of knowledge.

Consistent with Ball, other researchers found that a substantial portion of preservice teachers had difficulty in choosing the appropriate operation to solve multiplication and division word problems and many teachers' conceptions of division relied heavily on the domain of whole number and on the primitive partitive model (Graeber, Tirosh, & Glover, 1989; Tirosh & Graeber, 1990).

Post et al. (1991) described the mathematics knowledge profiles of about 220 mathematics teachers (Grades 4–6) in the domain of rational number concepts. Using the data from a test consisting of short answer items, a test requiring pedagogical explanations of solutions, and a structured interview, they concluded that "these results are quite disconcerting" (p. 186). A significant percentage of teachers missed one-half to two-thirds of the items, and in an extreme case, $1/3 \div 3$ was answered correctly only by 54 percent of the teachers.

Lee (1992) studied 42 prospective secondary teachers' understanding about the mathematical concept of limits in terms of three types of knowledge: subject matter knowledge, curriculum knowledge, and pedagogical content knowledge. The results of the study revealed that those prospective teachers' understanding of limit concept was more procedural oriented; there existed discrepancies between the subjects' concept definitions and their concept images of limit; and their misconceptions, difficulties, and errors were similar to those found in research studies on students.

Based on her doctoral dissertation, which investigated 152 preservice secondary teachers' subject matter knowledge and its interrelations with pedagogical content knowledge in the context of teaching the concept of function, Even (1989, 1993) reported that many of the subjects did not have a modern conception of function, very few could explain the importance and origin of the univalence requirement and appreciate the arbitrary nature of functions, and many chose to provide students with a rule to be followed without concern for understanding.

It is interesting to note that in addition to Even's dissertation (1989), there are at least four other dissertations investigating teachers' knowledge of the same mathematical topic: function (McGehee, 1990; Wilson, 1992; Bolte, 1993; Ebert, 1994). On the one hand, it perhaps reflects the fact that the conception of function plays a key role in school mathematics curriculum. On the other hand, it may explain that more discussion among different researchers is needed.

Baturo and Nason (1996) presented another picture of preservice teachers' knowledge in the domain of area measurement. Clinically interviewing 13 first-year primary education students in Australia with the instrument of eight area-measurement tasks, they found that the students' knowledge of area measurement was rather impoverished in nature. "Much of their substantive knowledge was incorrect, and/or incomplete, and often unconnected. The ability of the students to transfer from one form of representation to other forms of representations thus was very limited." Moreover, "their knowledge about the nature and discourse of mathematics and about mathematics culture and society was similar alarming".

Besides those investigating teachers' knowledge in specific domains, researchers have also examined teachers' general knowledge of a subject matter or of its pedagogy.

Thompson (1984) conducted case studies of three junior-high-school teachers to investigate their conceptions of mathematics and mathematics teaching. Through classroom observations, interviews, and written tasks, the research provided a general pattern of each subject's conceptions of mathematics and mathematics teaching. While the first teacher's teaching reflected a view of mathematics as a coherent collection of interrelated concepts and procedures, the second reflected more of a process-oriented approach than a content-oriented approach, and the third reflected a view of mathematics as prescriptive in nature and consisting of a static collection of facts, methods, and rules necessary for finding answers to specific tasks.

Eisenhart *et al.* (1993) did a case study of one student teacher, exploring her ideas and practices of teaching the sixth- and seventh-grade mathematics for understanding. They found that "although the subject believed in the importance of teaching for both procedural and conceptual knowledge, she was more confident in her arithmetic skills than she was in her conceptual knowledge, she had difficulty articulating how she would teach for conceptual knowledge, and she could not complete conceptual explanations for common topics in the elementary and middle school curriculum" (p. 17), and this limitation in knowledge base affected her ability to teach conceptual knowledge.

A similar study was done by Labouff (1996) to investigate mathematics teachers' conceptions of what it means to teach mathematics "for understanding". By mainly using a qualitative approach including interviews and observing teaching practices and analyzing the data collected from six seventh-grade mathematics teachers as the subjects, Labouff found that the majority of those teachers possessed little knowledge of expert conceptions of teaching for understanding and maintained a transmission model of teaching, even though all teachers had integrated into their practice teaching strategies associated with the reform movement in mathematics education which called for teaching for understanding.

In a study of elementary preservice teachers' knowledge and beliefs about science and mathematics, Stevens and Wenner (1996) tested 67 such undergraduate students using 30 questions in mathematics which were designed to assess students' conceptual understanding of mathematics including numbers and operation, measurement, geometry, data analysis, statistics and probability, and algebra and functions. Analyzing the scores of the subject, they concluded that there was a general weak knowledge base in mathematics (as well as science) among preservice teachers, and that the organization of knowledge at the conceptual level should be stressed in teacher education programs.

Overall, among the available studies investigating what knowledge teachers have, the majority have consistently revealed that teachers' knowledge is very insufficient in quantity and unsatisfactory in quality. Understandably, that is an essential reason why there have been so many reports and documents calling for improving teachers' knowledge. It is also one reason that it is valuable for this study to investigate how teachers developed their pedagogical knowledge, which is at the core of teachers' knowledge structure as revealed earlier.

How do Teachers Develop their Knowledge?

Compared to the above two issues, the last issue, how do teachers develop their knowledge, has been ignored for a long time. In fact, this issue has received little or no attention at all in many comprehensive studies and extensive literature reviews on teacher knowledge or teacher professional development (e.g., see Reynolds, 1989; Carter, 1990; Houston, Haberman, & Sikula, 1990; Aichele, 1994; Fenstermacher, 1994). Among the available research literature more or less related to the issue, a majority of studies have focused on some particular sources of teachers' knowledge such as attending a special preservice or inservice training program or even an educational course, and evaluated the influence of those sources on the development of teachers' knowledge.

Although not particularly focusing on teacher knowledge, Lortie's sociological study of school teachers drew wide attention to the influences of teachers' own experiences as school students on their perspectives of teaching (1975). According to his explanation, students on average spent about 13,000 hours (p. 61) in their school life closely observing their classroom teachers' teaching; this early and long-time "apprenticeship of observation" gave them a close-up and extended view of teaching. Moreover, despite the fact that this sort of knowledge about teaching was "gained from a limited vantage point and relying heavily on imagination", it was not fundamentally altered throughout the pedagogical training and was hard to change by their later work experience (Lortie, 1975, Chapter 3). While Lortie's analysis has been well known and seems plausible, I think his reason was essentially dialectical, and his empirical evidence, though helpful, was in a sense not strong.

Book, Byers, and Freeman (1983) did a survey of 473 entering teacher candidates at Michigan State University. Using a questionnaire which included questions about the sources of their professional knowledge, they found that these prospective teachers expected "on-the-job training" and "supervised teaching experiences" to be the most valuable sources, with the other four sources, clearly outdistanced by the former two, being courses in educational psychology, their own experiences as K-12 students, self-directed reading in education, and courses in the social-philosophical foundations of education, from high perceived importance to low. They also reported that nearly 80% of the preservice teachers had first-hand experiences with children as camp counselors, teacher aides, or Sunday school teachers, and most had "played teachers" in their childhood recreation. According to Book *et al.*'s analysis, this sort of experience was a main source of those preservice teachers' view of teaching as an extended form of parenting, and affected their (then) knowledge of teaching and expectation for education courses.

Unlike Lortie's analysis, Book *et al.*'s study left an interesting question open, that is, can teacher education programs change preservice teachers' belief, or give them new knowledge about teaching? If the answer is yes, then how?

Focusing on preservice teacher education and using case study methodology, Grossman (1988) investigated the influence of subject-specific teacher education coursework on the development of English teachers' pedagogical content knowledge. By analyzing the data from three beginning teachers with preservice training and three without preservice teaching training, she found that the two groups of teachers differed in their pedagogical content knowledge in important ways such as in their conceptions of teaching goals, content choosing, and concrete teaching strategies. She argued that pedagogical courses were helpful for the three beginning teachers to learn and develop their pedagogical content knowledge.

McDiarmid (1990) described a four-week field experience in which a group of prospective teachers in a teaching course were deliberately brought face-to-face with their tacit assumptions through encounters with negative numbers, third-graders, and an unconventional teacher. The researcher stated that some students openly resisted the implications of their experience, and while other students were willing to reconsider their understandings and beliefs, such changes might be superficial and short-lived. McDiarmid stressed that individual belief about teaching, learning, learners, subject matter knowledge, and context were already interwoven and hard to change by a single course.

It is interesting to note that although admitting that a ten-week mathematics methods course did have some impact on the preservice teachers in it, Ball (1989) raised the same concern as McDiarmid did about the extent to which this impact would continue. Moreover, a broader question that has been touched on much earlier but remains open is, will the effects of preservice training experiences at universities be washed out by school teaching experiences (e.g., see Zeichner & Tabachnick, 1981)?

Lappan and Theule-Lubienski (1994) described a study of 24 preservice mathematics teachers which was based on an intervention designed to understand what it took to help these preservice teachers confront their beliefs of mathematics and mathematics teaching and learning. They designed three mathematics courses, two methods courses (one before and one after student teaching), and seminars during student teaching. According to their report, after a year of the intervention in which students were helped to create a new vision of mathematics learning and teaching different from what students brought earlier, they in very powerful ways changed how the students perceived themselves as mathematics learners, but nearly half of them still held to their more traditional beliefs about what mathematics was important for elementary children and how to teach children mathematics. After the intervention in preservice education, they followed a subset of those students through their first three years of teaching. They concluded that disciplinary knowledge and a disposition to engage in mathematics inquiry or sense-making can be developed in an intervention like theirs, however, it is not enough to overcome the "deeply-held beliefs" about how young children should learn mathematics and what is important for them to know, which cannot be done solely in the preservice phase of teacher education.

Jones and Vesilind (1996) reported a study which investigated the changes of pedagogical knowledge of 23 preservice teachers in middle-grade teacher education during the senior year. Using data from concept maps, a card sorting task, and a structured interview, they found that student teachers reconstructed their pedagogical knowledge during the middle of student teaching, and their concepts of flexibility and planning for teaching changed rapidly. The students attributed these changes in knowledge organization primarily to student teaching experiences. The researchers concluded that "the influence of university course and cooperating teachers gave way to the influence of students and student-teacher interaction" (p. 113). From their study, they described student teaching "as a process of implementing prior knowledge about theory and methods, experiencing anomalies in this implementation, and, perhaps most importantly, reconstructing prior knowledge to account for experience and to create for oneself more coherent concepts about teaching" (p. 115).

In contrast, Shannon (1994) also did a similar study that investigated the effects of university educational coursework and clinical experience on the development of preservice teachers' knowledge. The result, which is totally different from Jones and Vesilind's, supported the contribution of professional coursework to the development of pedagogical knowledge, but showed limited support for the contribution of student teaching.

Foss and Kleinsasser (1996) provided another in-depth study investigating how 22 preservice teachers' beliefs and practice of teaching mathematics changed during a mathematics methods course of elementary teacher education. The students attended a 16-week course and taught three practice lessons in schools. Using both qualitative and quantitative data, Foss and Kleinsasser found that those preservice teachers' mathematics content knowledge and pedagogical content knowledge both remained constant throughout the semester course. Noticing the difficulty in changing preservice teachers' conceptions and influencing their development, and the inadequacy of current methods courses, they argued that "conceptions of mathematics teaching and learning must be moved to a conscious level with all involved in mathematics teacher education and teacher education research" (p. 441).

Compared to Foss and Kleinsasser's, it is interesting to note a similar study with different results reported by Langrall *et al.* (1996), which also assessed the influence of a one-semester elementary mathematics methods course on 71 preservice teachers' beliefs about teaching and their instructional practice. The methods course was intentionally designed to enhance pedagogical content knowledge, provide forums for reflection (an idea from Schön (1983)), and create opportunities for collaboration. Focusing on six of those students as case studies, the researchers found that the influence was successful and positive. They concluded that a mathematics methods program based on knowledge, collaborative support, and reflection may well have raised the consciousness of prospective teachers such that they could overtly affirm desired teaching actions and reject the undesired.

Researchers have also investigated, within different contexts, the influences of inservice training and professional experiences on teachers' knowledge. In his dissertation, Garoutte (1980) reported the effects of a special inservice training program, which lasted about one year and consisted of a total of 100 hours of workshop sessions, on the pedagogical knowledge of 100 elementary teachers. The 100 elementary teachers were administered a pre-test and a post-test called "The Test of Current Instructional Principles and Practices" twice. A control group of approximately the same number of elementary teachers was also administered the same post-test. Using a statistically significant test, Garoutte concluded that the inservice training program produced a positive change in teachers' pedagogical knowledge.

Feiman-Nemser and Parker (1990) examined conversations between four mentor teachers, two elementary and two secondary, and their four novice (first-year) teachers of different subjects including mathematics and English to discover how they discussed issues about the teaching and learning of content. The result revealed that mentors dealt with novices' understanding of subject matter very differently, including presenting subject matter directly, indirectly, assuming adequate subject matter knowledge, and ignoring it. They argued that subject matter should not be ignored in beginning teacher assistance programs; with experienced teachers' appropriate help, novice teachers can deepen their subject matter understanding, learn how to think about subject matter from the students' perspective, how to represent and present academic content, and how to organize students for the teaching and learning of subject matter. In short, novice teachers can develop their subject matter knowledge and pedagogical content knowledge from such programs.

Scholz (1995) presented a story somewhat different from Garoutte's and Feiman-Nemser and Parker's about the influence of inservice training programs on teachers' knowledge. The subjects of the study were eight inservice teachers who began teaching middle-school mathematics and participated in a two-year pilot professional development program. Utilizing data collected from those teachers via interview, questionnaire, and a unit work sample, Scholz found that those teachers had been taught mathematics in the early grades by worksheet, drill and practice, memorization, and flash cards, and their previous conception and knowledge structures regarding mathematics teaching were not changed significantly by the training program.

Jones (1997) compared 69 pedagogically trained and untrained English and science teachers' classroom performance in Barbados, West Indies, where there is no preservice training for teachers, and the formal training is provided for teachers with two or more years of teaching experience. Using statistical analysis, he found there were no significant differences in the total performance scores of these two groups of teachers, and their classroom teaching patterns were similar. He argued that it is probable "that the training programme is equipping teachers with skills which are so basic to successful teaching that untrained teachers quickly acquire those skills during the first or second year of teaching" (p. 183). The result raised a question pertinent to this study, that is, compared with formal teacher training, is teaching experience really a more important source of their knowledge of teaching?

Focusing on the "chronic problems" associated with teacher education, Lanier and Little provided a general picture of the influences of both preservice and inservice teacher education on teachers' learning to teach in their review of broad and diverse studies on teacher education. They described teacher education as "deintellectualization", and concluded that "prospective teachers find little intellectual challenge in their professional life", and "the world of elementary and secondary schools has not offered a more positive environment for learning to teach (than universities offering preservice training)... [T]he typical teaching experience of teachers in schools is noneducative at best and miseducative at worst" (Lanier & Little, 1986, p. 565). It seems to me their conclusion was too pessimistic, and ignored that there existed different practices of teacher education and focused too much on problems. Also, I do not think it is appropriate to treat "staff development" and "continuing teacher education" as synonymous terms. To me, teachers' conscious self-development such as self-learning and self-reflection is an extremely important part of teachers' development, which should not be generalized into inservice teacher education.

In general, there are several characteristics in the body of relevant research literature on how teachers develop their knowledge.

First, researchers have studied different kinds of sources for teachers to develop their knowledge, such as school experiences as students, preservice training, inservice training, and teaching experiences. They come to different conclusions about the extent to which each of those sources influenced the development of teachers' knowledge.

Second, almost all studies dealt with only one or sometimes two sources, mostly a preservice or inservice training program, for teachers to develop their knowledge. It seems obvious to me that, because of this common limitation, those studies can only reveal how important a source is from its absolute value to the development of teachers' knowledge. As teachers might have many other sources to develop their knowledge, it is unclear how important it is compared to other resources, namely, its relative value.

Third, along with the second feature, most studies are limited to a specific period of teachers' professional life, usually their preservice training period or the first several years of their teaching career. How teachers develop their knowledge during their whole professional life is still unclear.

Fourth, most studies are on a small scale and the subjects for the study were not randomly selected. While with a small number of subjects, the study can be more detailed and easier to conduct in terms of both research time and expenses, the findings are less representative and more dependant on specific contexts (such as different contents of a method course, or even different teachers of a methods course). This might be used to explain why some studies' findings are not consistent, or are even conflicting.

Last, almost all studies only indirectly or partially addressed the question of how teachers develop their knowledge. Overall, the available literature is not only scant, but fragmented, and systemic studies are virtually nonexistent. The concern that Griffin raised broadly about teachers' professional development — "there are so few conceptually sound and methodologically rigorous research studies related directly to staff development" (Griffin, 1983) — remains to be addressed.

Summary of the Review

To summarize, the review in this chapter reveals the following:

(1) Overall, knowledge has not been well conceptualized and consistently used across the studies on teacher knowledge, and especially, the epistemological background of the concept of knowledge has been largely ignored.
(2) Although different researchers proposed many different models to explain what knowledge teachers need in order to teach effectively, all of them listed pedagogical knowledge, along with subject matter knowledge, as an essential part of teachers' core knowledge.
(3) Among the available studies investigating what knowledge teachers have, the majority have consistently revealed that teachers' knowledge is very insufficient in quantity and unsatisfactory in quality.
(4) As to the question of how teachers develop their knowledge, almost all relevant studies only indirectly or partially addressed the question. School experiences as students, preservice training, inservice training, and teaching experiences are mostly treated as sources of teachers' knowledge, but the conclusions of those studies about the influences of those sources on teachers' knowledge are inconsistent within the same sources and fragmented across the different sources. There has been virtually no direct and systematic study focusing on the question, and new studies in this line are in high demand.

Chapter 3

A Conceptual Framework of the Study

The central question of the Chicago study is, "how do teachers develop their pedagogical knowledge?" To address this question, it is necessary at this stage to clarify some key constructs: knowledge, teachers' knowledge, teachers' pedagogical knowledge, and sources of teachers' pedagogical knowledge, and establish a conceptual framework to guide this study. This chapter is devoted to this task.

Knowledge

The discussion about the concept of knowledge in Chapter 2 has revealed that it is very difficult to give a precise definition of the term in epistemology, but this does not mean that knowledge can be used in an arbitrary way in research work; some kind of definition is necessary for implementing studies.

Before I give a definition of knowledge for this study, let me first clarify one thing about knowledge which I think is obvious but is very important to understand about the nature of knowledge, that is, any knowledge is inseparable from its subject and object. The subject is the holder of knowledge and the object is what knowledge is about.

In this study, I confine the subject of knowledge, as epistemologists usually do, to human beings (a person or a group of people), though some scholars, such as Russell and Ayer, have occasionally in their writing taken animals (e.g., a dog) as the subject of knowledge (e.g., Russell,

1948, p. 445; Ayer, 1956, p. 7; also see Campbell, 1988, p. 398). In other words, the study only deals with "human knowledge".

Unlike an object (e.g., a mountain), which can exist as an entity in the physical world whether or not human beings exist, knowledge of an object (the mountain) cannot exist independently of its subject, i.e., human beings. If there is no human being, there is no knowledge of the mountain. However, even if there are human beings, there is still no knowledge of the mountain until it is known by human beings. As to the object of knowledge, a subject's knowledge is always about something. When people ask "do you know", it always implies, "do you know that (something)". When people say "someone lacks knowledge", it essentially means "someone lacks knowledge about something".

In other words, there is no "absolute" knowledge; the subject and the object of a piece of knowledge precede the knowledge itself. That is, any knowledge must exist in someone and be about something; and when we talk about "knowledge", we in essence mean "a subject's knowledge of an object".

Therefore, it seems to me that the logical starting point to understand the concept of "knowledge" is not "what knowledge is" but "what a subject's knowledge of an object is". Here is my basic definition from the perspective of epistemology for this study:

A subject's knowledge of an object is a mental result of certain interaction of the subject and the object.

Some explanations and restrictions are necessary for the definition. In the definition:

"A subject" is a person, or a group of people, but not an animal, or a machine like a computer or robot, which is excluded in this study, though in a different sense we can say a dog knows his master, and still in a different sense a computer knows how to conduct some operations, and a robot like "Deep Blue", a chess-playing computer developed by IBM, knows how to play chess.

"An object" can be anything, such as a place, a thing, a method, a procedure, a person, an activity, a piece of knowledge previously known by the subject or others, a discipline, and the like, or any combination of them.

"Certain interaction" can be a process of the subject's acquainting, observing, experiencing, reflecting, reasoning, thinking, and the like, with, on, or about the object,[1] but not purely guessing, arbitrarily imaging, or nonsense dreaming.

I use "interaction" in order to stress that sometimes an object is not still, and the process of a subject's knowing of the object is dynamic.

There are different kinds of interactions from which people can obtain their knowledge. The interaction can be direct or indirect. For example, a child can know $4+3=7$ by himself by putting four marbles and three marbles together, then counting the total; or by just listening to his teacher's explanation and memorizing the result. The interaction can also be mainly physical or purely mental. For example, a person can know an object is very heavy just by lifting it, and a student can know the solution to the equation $\sin^2 x - \cos^2 x = 1$ by mathematical reasoning in his mind. Moreover, it can be instant or very long. For example, an infant can know his mother is around instantly from her voice, and it may take a mother an hour to know why her baby is crying. Last, it can be an active or passive process for the subject. For example, a layman can passively memorize a scientific result by watching a TV program, and a scientist can actively find or confirm the result by doing experiments. How different kinds of interactions affect the origination, evolution, and development of knowledge should be a central issue in the theory of knowledge.

"A mental result" is a subject's mental or cognitive achievement of the interaction, including belief, memory, or understanding,[2] but not a mental emotion, tendency, or inclination, etc. A mental result is not necessarily expressible in words, or even sometimes noted by the knower (namely, sometimes a knower does not know what he actually knows). Thus a mental result can be "implicit" or "explicit".

---

[1] According to Laird, Dewey described "knowledge" as "the result of thought" (see Laird, 1930, pp. 101–106). It is obvious that the definition here is consistent with, but broader than, Dewey's.

[2] Waddington recently claimed that understanding is connected knowledge, with which I basically agree. However, he did not explain what knowledge is (Waddington, 1995).

From the definition, we can term the subject "the knower", the object "the known", and the interaction of the subject and the object "the knowing". Briefly, it can be pictured as below:

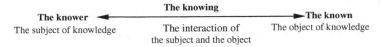

The knower	The knowing	The known
The subject of knowledge	The interaction of the subject and the object	The object of knowledge

Being warned of the difficulty, in classical writings of epistemology, in defining "knowledge", I do not claim the above definition is perfect, because it still contains ambiguity; and, to me, although "a subject" and "an object" in the definition are clear, "certain interaction" is only described, but not precisely defined, and so is "a mental result". However, the above definition does prescribe that knowledge is "something", not anything, and enables us to clarify several things which I think are highly desirable.

First, knowledge defined in this definition includes all knowledge traditionally defined in epistemology as "justified true belief" or its expanded version as "objectively grounded belief", because "to be justified true" or "to be objectively grounded" necessitates some kind of "interaction". However, this definition contains knowledge more than just "belief". From the definition, it is easy to recognize that knowledge can be knowing that (e.g., a belief), knowing things (e.g., a memory), and knowing how (e.g., understanding).

Second, this definition emphasizes that knowledge, whether it is belief, memory, or understanding, can be "true" or "false", as long as it results from "certain interaction". In contrast, without such interaction, a mental result (e.g., a belief or a memory) is not knowledge, whether it is true or not. For example, superstitious belief is a belief (Quinton, 1967), but not knowledge, even it can become true; a person may remember a nonsense dream, but what he remembers about something from the dream is not knowledge.

There are two main reasons for me not to exclude "false knowledge" in the definition as many epistemologists traditionally have done. One is that, unlike mathematical knowledge, many times it is impossible or meaningless to tell if a teacher's knowledge, for example, of a teaching method, is totally wrong or right (also see Orton, 1993; Donmoyer, Imber, & Scheurich, 1995, pp. 3–6); often we can only tell if his knowledge of that method is quite good or not so good, and even this is based on our own judgment. The other reason is that, even for teachers' false knowledge, I think it is still worthwhile to know from what sources teachers developed such kind of knowledge.

Third, in light of the definition, knowledge is a result (mental achievement) of a process (interaction), not the process itself. From this viewpoint, knowledge is not experience itself or any other kind of activity, although some researchers have either unintentionally or intentionally mixed them up (e.g., personal experience and personal knowledge) up. This is to me most apparent in the well-publicized studies of teachers' personal practical knowledge, in which teachers' knowledge is typically treated as "the sum total of the teachers' experiences" (e.g., see Elbaz, 1983; Johnson, 1989; Clandinin *et al.*, 1993; Connelly, Clandinin, & He, 1996).

Fourth, the definition reveals that knowledge is in the mind ("mental result"). It is not in the "body" (e.g., see Connelly & Clandinin, 1988, p. 25). Also, I think it is not appropriate to treat "skill" as a kind of knowledge (e.g., see Alexander *et al.*, 1991), for skill is often (though not always) a person's ability to do something. In other words, knowledge deals with what people know, but skill deals with what people can do; even though they are closely related, and some researchers of teacher knowledge have claimed that skills teachers use are inseparable from the knowledge they hold (e.g., see Corrie, 1997).

Last, some researchers have raised the question: what is the connection between the knower and the known? (e.g., see Greene, 1994, p. 425). According to the definition here, the basic connection is "interaction", that is, a person's knowledge about an object comes from his interaction with the object. In addition, different levels of interaction may produce different degrees of knowledge.

It is important to note there are two different kinds of the knower: an individual or a group of individuals. Correspondingly, there are two kinds of human knowledge: individual knowledge or collective knowledge. Ultimately, collective knowledge comes from individual knowledge. For example, human knowledge of calculus in mathematics and that of relativity in physics were first individual knowledge, when the former was initially invented by Newton and Leibniz in the 17th century, and the latter by Einstein in the early 20th century. Now they have become collective knowledge, widely learned and known by many high-school and university students. This explains the knower of knowledge is the holder of knowledge, not necessarily the ultimate producer of the knowledge. When talking about the development of knowledge, we need to be aware knowledge can not only be produced, but also transmitted and learned. To many people, learning from others through various means is the most important way to develop their knowledge.[3]

Teachers' Knowledge

"Teacher knowledge" and "teachers' knowledge" are often used by researchers without distinction. In this study, I also treat them as synonymous words. From the perspective of the knower and the known of knowledge, researchers have employed "teacher knowledge" to refer to the following different meanings:

(1) The knower: teachers; the known: what teachers know. Namely, the term means "knowledge possessed by teachers".
(2) The knower: researchers; the known: teachers. Namely, the term means "knowledge about teachers".
(3) The knower: researchers and/or teachers; the known: teaching. Namely, the term means "knowledge related to teacher (about teaching)" (e.g., see Fenstermacher, 1994).

This study uses the first meaning as its definition of "teacher knowledge", as I think it is quite different from, and basically should not be mixed up with, the other two meanings. In other words, in "teachers' knowledge" or "teacher knowledge", teachers are the knower, not the known, though the study of "teacher knowledge" will produce knowledge about teachers.

Buchmann pointed out that teacher knowledge implies that what is known is special to teachers as a group (which I agree), but the term "teaching knowledge" allows considering knowledge related to the activities

---

[3]This fact is often neglected. For instance, Mao (1971) claimed, in his famous lecture of 1963, "Where do correct ideas come from?", that human knowledge can only be from social practice. To me, it seems obvious that Mao was talking about the ultimate origin of knowledge.

of teaching while leaving that question open (Buchmann, 1987, p. 152). She preferred using "teaching knowledge" instead of "teacher knowledge". However, to me, those two terms are not just different, but also not replaceable with one another. In "teacher knowledge", the knower is teachers, and the known is what teachers know as teachers, which is related to teaching, but not just about teaching; in "teaching knowledge", however, the knower is unknown, and it can be teachers, teacher educators, researchers, or even laymen, while the known is relatively clear, that is what is known about teaching.

Not only is teachers' knowledge different from the general public's, teacher educators' or educational researchers' knowledge, it is, as Gilbert *et al.* (1987) pointed out, also not the same as the knowledge of other school educators such as those in administration and counseling.[4] By "teachers' knowledge", this study only focuses on knowledge possessed by teachers specifically as teachers, not as general people in society, or as other kinds of educators in school, such as department chairs or counselors, though some subjects of this study did hold such positions when the empirical data was collected in this study (see details in Chapter 4).

Researchers have argued that the knowledge which guides the practice of teachers is much more than an accumulation of propositional facts (e.g., see Rubin, 1989; Johnston, 1992). Moreover, Orton (1993) pointed out two issues with the study of teacher knowledge: first, teacher knowledge appears to be primarily a form of knowing *how*, and hence it is often "tacit". According to him, the most credible justification for a teacher's knowledge is that the teacher can do something in the classroom. Second, teacher knowledge is deeply dependent on particular times, places, and contexts, and lacks the character of knowledge in mathematics, physics, or even psychology, and hence is "situated". Thus it is hard to formulate criteria which can be used to explain how a piece or instance of teacher knowledge might be justified.

In my view, following the definition of knowledge aforesaid, "teachers' knowledge" means "what teachers know (as teachers)". It includes: teachers' "beliefs", such as which strategy is most effective to

---

[4]Actually, like the study of knowledge base for teachers, there has been research about the knowledge base for educational administrators (e.g., see Donmoyer *et al.*, 1995).

teach a special topic; "memory", e.g., what is the structure of a textbook; and "understanding", for instance, how to use a teaching method, and so on.

Teachers' knowledge can be of various forms or types. Noticing there are other distinctions made by epistemologists with overlaps about different kinds of knowledge, I think, at this point, it suffices to mention that there are three essential kinds of teachers' knowledge: "knowing things", "knowing that", and "knowing how"; and their knowledge can be either "tacit" or "explicit".

Teachers' Pedagogical Knowledge

Because teachers' pedagogical knowledge is literally equal to teachers' knowledge of pedagogy, at this point, it seems helpful first to describe briefly what pedagogy usually means.

In its broadest meaning, pedagogy is used to mean "the study of education", or sometimes "education" itself (e.g., see Good, 1945; Barnard & Lauwerys, 1963); in its narrowest, it refers to "the study of the ways of teaching", or simply "teaching methods" (also see Rowntree, 1981). Between them, it is also commonly defined as "a science of teaching embodying both curriculum and methodology", or a little wider as "a science or art of teaching" (e.g., see Monroe, 1913; Simon, 1981; Lawton & Gordon, 1993).[5] Therefore, despite the differences people mean by the word of pedagogy, all include "teaching" and "teaching methods" as part of its meaning. To me, the core meaning of pedagogy is about (not just related to) teaching and teaching methods.

It is understandable that various knowers, such as teacher educators, researchers of pedagogy, or even people in general, might have some kind and amount of pedagogical knowledge. However, this study only deals with teachers' pedagogical knowledge. Needless to say, it is more important for teachers to have adequate pedagogical knowledge; the width, the depth, as

---

[5]In many language dictionaries, "pedagogy" is also explained to mean "the function, profession, or practice of a teacher" (e.g., see *The Oxford English Dictionary* (1961); *Webster's New Universal Unabridged Dictionary* (2nd edition, 1983); & *Random House Unabridged Dictionary* (2nd edition, 1993)). However, this meaning of "pedagogy" is rarely used in educational circles; see the references in this paragraph.

well as the purpose of pedagogical knowledge possessed by teachers and that by other knowers are different.

As the review in Chapter 2 has revealed, although it is no longer controversial that, like subject matter knowledge, pedagogical knowledge is a key component of teachers' professional knowledge to good teaching, there are differences among researchers' uses of the meaning of pedagogical knowledge, just like the meaning of pedagogy.

Some researchers have applied it only to teachers' knowledge of general teaching procedures such as effective strategies for planning, classroom routines, behavior management techniques, classroom organizational procedures, and motivational techniques (e.g., see Fennema & Franke, 1992, p. 162); others have used it to include both generic teaching methods and theories applicable to any subject area and pedagogical content knowledge (e.g., see Colton & Sparks-Langer, 1993). Still others have employed the term more broadly to include knowledge of students, curriculum, planning, instruction, and evaluation (e.g., see Jones & Vesilind, 1996).

In the subject of mathematics, NCTM's *Professional Standards for Teaching Mathematics* (1991) stressed that, "Mathematics pedagogy focuses on the ways in which teachers help their students come to understand and be able to do and use mathematics" (p. 151). The *Standards* identified the following five components of teachers' pedagogical knowledge under the rubric of "knowing mathematics pedagogy", which it claims are essential to quality teaching:

A. knowledge of instructional materials and resources, including technology

Teachers are responsible for posing worthwhile mathematical tasks. They may choose already developed tasks or may develop their own tasks to focus students' mathematical learning. To do so, they often rely on a variety of instructional materials and resources, including problem booklets, concrete materials, textbooks, computer software, calculators, and so on. Teachers need a well-developed framework for identifying and assessing instructional materials and technological tools, and for learning to use those resources effectively in their instruction (NCTM, 1991, p. 151).

B. knowledge of ways to represent mathematics concepts and procedures

Modeling mathematical ideas through the use of representations (concrete, visual, graphical, symbolic) is central to the teaching of mathematics. Teachers need a rich, deep knowledge of the variety of ways mathematical concepts and procedures may be modeled, understanding both the mathematical and developmental advantages and disadvantages in making selections among the various models. In addition, teachers need to be able to translate within and between modes of representations in order to make mathematical ideas meaningful for students (NCTM, 1991, p. 151).

C. knowledge of instructional strategies and classroom organizational models

Criticizing the traditional teaching approach to mathematics — the "tell, show, and do" model — the *Standards* stress that teachers need to employ alternative forms of instruction that permit students to build their repertoire of mathematical knowledge and their ability for posing, constructing, exploring, solving, and justifying mathematical problems and concepts. Promising models for such instruction are all highly interactive. In such models, teachers both model and elicit mathematical discourse by asking questions, following leads, and conjecturing rather than presenting faultless product (NCTM, 1991, p. 152).

D. knowledge of ways to promote discourse and foster a sense of mathematical community

Teachers need to focus on creating learning environments that encourage students' questions and deliberations — environments in which the students and teachers are engaged with one another's thinking and function as members of a highly teacher–student and student–student interactive mathematical community. They also need to employ strategies that will help them develop the participation essential to engaging students in mathematics, such as student group work and placing responsibility on students (NCTM, 1991, p. 152).

E. knowledge of means for assessing student understanding of mathematics

It is pointed out that teachers need to know ways to assess how students learn and what they are able to accomplish in order to deliver their teaching more effectively. Assessment should focus on addressing students' development of mathematical power. Teachers need to align assessment with instructional goals, consider their purposes in assessment, and understand the issues surrounding assessment in many aspects. As teachers' experiences with methods of assessment are often limited to the more traditional "testing and measurement" strategies provided through a preservice course, there is a strong need to integrate the understanding and use of alternative methods of assessment as an on-going topic throughout teachers' educational life (NCTM, 1991, p. 153).

Comparing the above five components to those proposed by different researchers reviewed earlier, we can see that, roughly, Part A corresponds to "curricular knowledge", Part B to "pedagogical content knowledge", and Part C to "general pedagogical knowledge". Parts D and E are usually not separately listed from the others by most researchers, and they, especially Part D, can be partially integrated into "pedagogical content knowledge" and partially into "general pedagogical knowledge". Moreover, Part A contains teachers' knowledge of technology, which, as the previous review revealed, has not received enough attention from many researchers but is important to teaching mathematics, especially in new reformed mathematics curricula.

In this study, "teachers' pedagogical knowledge" is generally defined as "what teachers know about how to teach mathematics". It includes teachers' knowledge of both "curriculum and methodology (of teaching)".[6] More specifically, it consists of the first three components of pedagogical

---

[6]That is, "how to teach mathematics" can be divided into "how to use curriculum to teach mathematics" and "how to use teaching methods to teach mathematics". Also see the definitions of "pedagogy" aforementioned in this section.

knowledge in NCTM's definition, which I term correspondingly as follows:

Pedagogical Curricular Knowledge: knowledge of instructional materials and resources, including technology;

Pedagogical Content Knowledge: knowledge of ways to represent mathematics concepts and procedures; and

Pedagogical Instructional Knowledge: knowledge of instructional strategies and classroom organizational models.

For brevity, I will hereafter use PCrK to denote "pedagogical curricular knowledge", use PCnK to denote "pedagogical content knowledge", and use PIK to denote "pedagogical instructional knowledge".

Though not absolutely, I think in some sense we can say that PCrK is more about *knowing things*, for instance, familiarity with the hierarchical structure of teaching contents, awareness of the resources available for teaching a special topic, and understanding of the features of the textbooks being used. PCnK is more about *knowing how*, such as knowledge of, in mathematics, how to represent the concept of function,[7] and of how to introduce students to solving a general quadratic equation. PIK is more about *knowing that* and *knowing how*, for example, what is and how to use cooperative learning in the classroom.

The main reason for this study focusing on the three components is that they have been better defined and widely accepted in the community of mathematics education, and I believe they are the core of mathematics teachers' pedagogical knowledge, as all of them are directly related to the way teachers conduct mathematics teaching in the classroom. As to the last two components in NCTM's definition, to a great degree how to promote discourse and foster a sense of mathematical community and how to assess students' understanding of mathematics in the classroom, as mentioned above, are essential parts of teachers' PCnK and teachers' knowledge of general strategies and classroom organizational models for teaching mathematics. For example, cooperative learning or group discussion is a general teaching

---

[7] Note that according to the analysis earlier, in this study, knowledge of "function" itself is not PCnK, rather it is subject matter knowledge.

strategy, however, in mathematics classrooms, it is often used to promote mathematical discourse and create a sense of mathematical community.

Sources of Teachers' Pedagogical Knowledge

In the literature on teacher knowledge, some researchers have used the term "source" for two different meanings without making a necessary distinction: means by which they developed their knowledge, and something (often prior knowledge) teachers acted upon within their minds to develop the knowledge. For example, Grossman (1988, 1991) listed three sources of PCnK: apprenticeship of observation, teacher education, and classroom experience, for teachers can acquire their pedagogical knowledge from their school experience as students, teacher-training courses, and actual teaching experience. Meanwhile, she also listed disciplinary (content, subject matter) knowledge as a source of teachers' PCnK because teachers' knowledge of a discipline can inform their decisions about choosing content and the development of PCnK (also see Shulman, 1987; Even & Tirosh, 1995). To me, the first three bear a different meaning of "source" from the last one in a substantial way and should not be juxtaposed as the same type. To make the distinction, I would consider the former as "macro" level and the latter as "micro" level.

The following examples explain the difference of the two meanings of "source". Suppose a traveler comes to believe that "the shortest way for airplanes to fly from Chicago to Shanghai is over Alaska, not over Hawaii" (Knowledge A), after his own reasoning from the fact he knows that "the shortest distance between two points on a sphere is found in the great circle of the sphere containing those two points" (Knowledge B). It will be natural to say that his knowledge of the shortest way from Chicago to Shanghai (Knowledge A) is from his knowledge of spherical trigonometry (Knowledge B), or in other words, his knowledge of spherical trigonometry (Knowledge B) is a source of his knowledge of the shortest way (Knowledge A).

However, suppose a pilot does not have knowledge of spherical trigonometry (Knowledge B), and he comes to know that it is faster to fly from Chicago to Shanghai over Alaska than over Hawaii after he flew both routes or after being told in a training course, it will also be common to say that his knowledge about the shorter route is from his experience or his training course; in other words, his experience or training is the source of his knowledge about the shorter route. If we use this viewpoint of "source" (at macro level) to look at the traveler's situation, then the traveler's Knowledge A resulted from his own reasoning using his Knowledge B. That is, his own reasoning itself, regardless how much his prior knowledge was utilized (at micro level) in this process, is the source of his Knowledge A.

Considering the definition of knowledge as a mental result of certain interaction of the subject and the object, I believe both of the above meanings of "source" are about "interaction".[8] However, when we seek the source of a piece of knowledge at the macro level, we want to know what "interaction", *across* all possible interactions, results in the knowledge; when at the micro level, we want to know, *within* an "interaction", what is used or acted upon to obtain the knowledge.

In this study, I only use the macro meaning for the word "source", i.e., the means by which teachers develop their knowledge, for I think, as an empirical study, it is much more useful and practical to know this kind of source in order to find ways to improve teachers' knowledge.

According to the literature review in Chapter 2 and my own analysis, which is partly from my personal experience both as a mathematics teacher in three schools and a teacher educator at a teacher-training college for about ten years, I establish the following framework to investigate the sources from which teachers (of mathematics) develop their pedagogical knowledge.

Briefly speaking, the framework has three major components: teachers' experience as learners before accepting formal preservice training, teachers' preservice training experience, and teachers' inservice experience; each of which includes several sub-components, as shown in Figure 3.1 below.

Experience as learner. By "experience as learner", I mainly mean teachers' experiences as learners before accepting formal preservice

---

[8]Recently, researchers also refer "source" to the knower, that is, who produced the knowledge about teaching. The focus has been on "researchers" or "practitioners (teachers)" as the producer of the knowledge (e.g., see Cochran-Smith & Lytle, 1993; Fenstermacher, 1994). This meaning of source is irrelevant to this study.

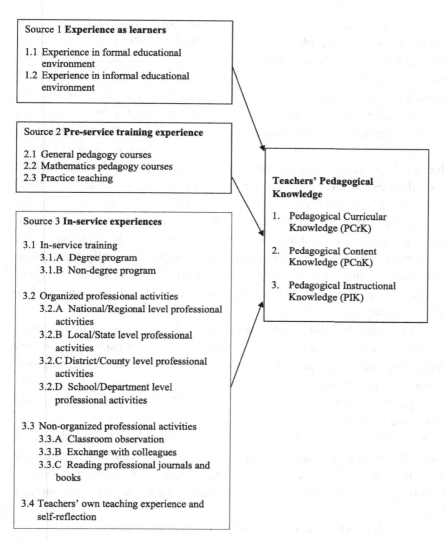

Figure 3.1 A framework to investigate the sources of teachers' pedagogical knowledge

training, which obviously happened mostly in formal educational environments, i.e. school, but could also happen in an informal educational environment, such as family and other everyday life situations. The previous review revealed that some researchers like Lortie (1975) argue that teachers' experiences as school students or "apprenticeship of observation", is an extremely important source of their knowledge of how to teach. Moreover, I believe the exposure of a person to informal educational environments, such as listening to their mother's storytelling and watching how an adult explained the answers to a question, can also bestow an individual with explicit or implicit memory about how to deliver a story, about how to communicate with listeners, and other knowledge useful for teaching, regardless of how superficial the knowledge is (also see Jackson, 1986, Chapter 1).

People who claim that teachers teach the way they were taught imply that teachers' pedagogical knowledge, which they believe mainly comes from their experience as school students, is PIK at least, or both teachers' PCnK and PIK at most. However, I want to add that this experience can also be a source of teachers' PCrK. This is particularly true when teachers teach the same textbooks (or of a revised version) as they learned in school. This phenomenon is not as rare as it might be thought, especially for popular textbooks in curriculum-decentralized countries such as the US and UK, and for national standard textbooks in curriculum-centralized countries like China and Japan.

Preservice training experience. There is no question that it is an essential purpose of preservice training to provide prospective teachers with adequate pedagogical knowledge and skills. As revealed in a previous review, "mathematics pedagogy courses" (or "methods courses") and "practice teaching" (or "student teaching") have been described and discussed by many researchers (e.g., Grossman, 1988; McDiarmid, 1990; Foss & Kleinsasser, 1996) as sources of teachers' pedagogical knowledge. "General pedagogy (educational) courses" is separately added here as I think students should be expected to learn PIK from the courses, which is not specifically for teaching mathematics or any other subject matters.

Inservice experience. Teachers' inservice experience herein means their professional experience after formally becoming school teachers. In Figure 3.1, it is divided into four sub-components. The first is "inservice training" experience by which teachers received formal and purposeful professional training, including graduate degree programs, and non-degree programs such as a summer course that specifically introduces teachers to a new teaching strategy (e.g., cooperative learning). Studies reviewed in Chapter 2 have shown that normal (non-degree) inservice training programs can be influential or insignificant for teachers to develop their pedagogical knowledge (e.g., Garoutte, 1980; Scholz, 1995).

The second sub-component, as listed in Figure 3.1, is teachers' experience of attending "organized professional activities". By "professional activities" I exclude those specifically designed for professional training, which have been taken into account above. "Organized professional activities" are those organized by a certain organization, such as a general conference, seminar, workshops, etc., which can be at the national/regional level, local/state level, district/county level, and school/departmental level. Although there seems little empirical research available investigating how much teachers benefited in terms of their pedagogical knowledge by attending such professional activities, it is reasonable to expect that these professional activities could contribute, more or less, to the development of teachers' knowledge.

The third is teachers' "non-organized professional activities", which are often more casual and can occur within teachers' everyday professional life. Under that I list "classroom observation", "informal (daily) exchanges with colleagues",[9] and "reading professional journals and books", for I believe they could be helpful for teachers to acquire new ideas about mathematics teaching, though those factors have been largely ignored in the literature on teacher knowledge.

The last sub-component listed in Figure 3.1 is "teachers' own teaching experience and self-reflection". It is self-evident that teachers can develop knowledge of teaching from their own teaching experience. Not only can teaching experiences (practices) reinforce or consolidate their previous knowledge which has proven to be right or workable, and correct or modify their previous knowledge which has proven to be wrong or unworkable, it can also provide important (though not necessarily sole) opportunities for teachers to acquire or produce many kinds of new knowledge (including those with characteristics of "situated knowledge" and "tacit knowledge"), such as what is the structure of a textbook (PCrK), what is the most difficult thing for students to understand in a lesson (PCnK), and what general teaching strategy works best for involving students (PIK). As a matter

---

[9]"Informal" is added here to contrast with "formal exchange with colleagues" (e.g., in professional activities organized in school).

of fact, numerous researchers such as Clandinin and Connelly (see the review earlier) and Britzman (1991) have closely dealt with how teachers' own teaching experience contributes to the development of their knowledge of teaching in their studies. In addition, by "self-reflection" herein I specifically mean teachers' own reflection which is originated from their experience, and is on their experience. Some researchers of mathematics education have conceptualized "reflection" as "the set of process (of evaluating one's own teaching) that enables a professional to learn from experience" (Brown & Borko, 1992, p. 212), or "the act of making meaning of one's experiences" (McLymont & da Costa, 1998; also see broader discussions in Schön, 1983, 1991). The reason for adding "self-reflection" is mainly that I believe what teachers can learn from their own teaching experience will be very limited both in quantity and quality without necessary self-reflection on such experience.

Below in this study, "teachers' own teaching experiences" is sometimes alternatively called "teachers' own teaching practices" and "self-reflection" is abbreviated as "reflection" in terms of this source.

I strongly doubt the common assumption that teachers learn their pedagogical knowledge mainly through their experiences as students, mentioned before. Instead I believe that teachers can learn pedagogical knowledge from all the sources framed above. For example, teachers might use the way they were taught (Source 1.1), they might learn how to use special computer software or calculators (e.g., TI-82) to teach mathematics from their methods course in preservice training (Source 2.2), and also they might learn specific teaching methods from attending a national meeting such as the NCTM annual meeting (Source 3.2.A) or from reading a professional book (Source 3.3.C). However, to me, how those sources comparatively contribute to the development of each component of teachers' pedagogical knowledge cannot be answered until this study is done with adequate empirical evidence from teachers.

Summary

In this study, knowledge is understood from the dynamic relation of the knower (the subject of knowledge), the known (the object of knowledge), and the knowing (the interaction of the subject and the object). A subject's knowledge of an object is defined as a mental result of certain interaction of the subject and the object.

From this definition, teachers' pedagogical knowledge is then defined as what teachers know about how to teach, which includes teachers' knowledge of both curriculum and methodology of teaching. More specifically, in mathematics, teachers' pedagogical knowledge consists of three components which are based on the NCTM *Standards*: PCrK — knowledge of instructional materials and resources, including technology; PCnK — knowledge of ways to represent mathematics concepts and procedures; and PIK — knowledge of instructional strategies and classroom organizational models.

To investigate how teachers develop their pedagogical knowledge, I have established a framework which is based on the previous review and my own experience and analysis. The framework presents possible sources for teachers to develop their pedagogical knowledge. It has three major sources: experience as learner, preservice training experience, and inservice experience, each of which includes a number of sub-components. Some of them have been well known yet others have been less well known and often ignored in previous studies. A brief rationale for including those sources was given above and all of the sources will be taken into account one way or another in this study.

To end this chapter, I would like to point out that, as we can see from the previous review, each of the three components of pedagogical knowledge, combined with each of the sources, could be important and broad enough to serve as a base for a specific study. The decision to study them together, though it may lead to a less detailed picture of some aspects, is made so that a more complete and systematic picture of how teachers develop their pedagogical knowledge can be illustrated, the interrelations among the different sources can be discerned, and how different resources contribute to the development of teachers' pedagogical knowledge can be compared, which are the main purposes of this study.

Chapter 4

Research Design and Procedures

The purpose of this study is to investigate sources from which teachers develop their pedagogical knowledge. Based upon the review in Chapter 2 and the conceptual framework discussed in the last chapter, a multiple-stage methodology was designed for the purpose of this study.

This chapter describes in detail the methodological issues of the study, including the targeted population and sample, research instruments, data collection procedure, methods employed to process and analyze the data, and the strengths and limitations of the methodology.

Population and Sample

Schools

In a strict sense, the school population of this study consists of the 25 best public high schools, making up 12.9% of all the public high schools in the metropolitan area of Chicago, Illinois. By "the best schools" I specifically mean the schools whose students' average scores in the 1996 Illinois Goal Assessment Program (IGAP) mathematics test for the tenth grade were at the top level of the 194 public high schools in the Chicago area, which includes Chicago city and its six suburban counties: Cook, DuPage, Lake, Will, Kane, and McHenry.

IGAP tests are state-required exams for all public schools in the state of Illinois. The school subjects included in the tests are reading, mathematics, science, social science, and writing, which are given to different grade students from elementary to high schools. The mathematics tests are given to the third-grade, sixth-grade, eighth-grade, and tenth-grade students on a 0–500 scale. Since being introduced in 1989, IGAP test scores have been widely used by the educational community in Illinois as well as by the media and the public as a criterion in determining the excellence of schools. Moreover, it has been shown in recent years that the rank of top schools in terms of students' average scores in IGAP tests is very stable.

For this study, all of the 25 schools have been identified using the database of "Illinois School Report Cards", which can be found in the on-line web sites of "Chicago Tribune" or "Chicago Sun-Times", two major newspapers published in the Chicago area. The tenth-grade students' average scores of the mathematics test in those 25 schools ranged from 322 to 375 with the mean of 340.6 and the median of 337, all significantly higher than the state average of 262.

The sample of schools consists of three schools. A stratified random sampling method was utilized so that the sample could be more representative of those 25 high-performing schools than a simple random sample. Namely, one school was selected randomly from the best eight schools in terms of the IGAP average score, another from the second best eight schools, and the other from the remainder of the 25-school population.

Teachers

The subjects of this study are all the mathematics teachers who were at the time teaching regular mathematics classes in the three sample schools either full time or, in a very few cases, part time (e.g., a person who also taught other subject matters, or had other responsibilities in the schools such as being the chair of the mathematics department). A few staff in the mathematics departments who were tutors, mathematics laboratory instructors, and administrative assistants were excluded. In total, there were 77 mathematics teachers in those three schools for this study.

All of these mathematics teachers were asked to complete a questionnaire. A subsample consisting of nine teachers out of the 77 teachers were each observed twice for their actual teaching; and 12 teachers, the nine observed and three math chairs, were interviewed. The subsample of nine teachers for the classroom observation and interview comprised three teachers in each school, excluding the chair of the mathematics department, with one randomly selected from teachers with 0–5 years' teaching experience, one from those with 6–15 years' teaching experience, and the other 16 or more years' teaching experience (see more details in the section on data collection below).

The size of the sample of this study was decided according to both the purpose of the study and the possibility of conducting such a study in terms of both time and financial expenses. The reason that the school population consisted of only high-performing schools was not only the fact that a large population with a relatively small size of the sample would make the findings from the sample less representative, but also, understandably, the practice of teachers' professional development in high-performing schools may contain more valuable information for us to find effective ways to pursue teachers' professional development. Also, the stratified sample of teachers for classroom observation and interview was used so that teachers with various amounts of teaching experience could be included in those stages and the possible difference of the sources of their pedagogical knowledge could be detected.

Instruments

The original data for this study were collected by applying three main research instruments: a questionnaire, classroom observation, and interviews following the observation.

Questionnaire

The questionnaire consisting of 22 questions can be found in Appendix 1A. General principles about wording and formatting for questionnaire construction (e.g., see Berdie, Anderson, & Niebuhr, 1986; Fink, 1995) were used to guide the design of all questions, and a number of experts as well as high-school mathematics teachers were consulted technically to make sure the questions could be answered clearly.

Conceptually, the design of the whole questionnaire is based on the framework discussed in Chapter 3. Generally speaking, Questions 1–4 are designed to provide background information that is needed to understand and analyze the teachers' answers to the questionnaire; Questions 5–15 focus on how teachers' specific experiences contributed to the development

Table 4.1 Distribution of the questions in the questionnaire in terms of the components of teachers' pedagogical knowledge and the knowledge sources

	Pedagogical curricular knowledge	Pedagogical content knowledge	Pedagogical instructional knowledge	Unspecified pedagogical knowledge	Total
Experiences as students	5, 16, 17, 18	—	5, 19	—	5
Preservice training	7, 16, 17, 18	7	7, 19	6, 8, 9	8
Inservice experience	12, 16, 17, 18	12, 22	12, 19	10, 11, 13, 14, 15	11
Unspecified sources (open-ended)	—	20, 21	—	—	2
Total	6	5	4	8	—

Note: Questions 1–4 are for obtaining background information needed to understand and analyze the teachers' answers to the questionnaire.

of their pedagogical knowledge (e.g., Question 9: "Overall, how useful was your preservice training in enhancing your knowledge of how to teach mathematics?"); in contrast, Questions 16–22 focus on how teachers' specific pedagogical knowledge was developed from their different experiences (e.g., Question 16b: "How much did the following sources contribute to your knowledge of how to use technology for teaching math?"). See details in Appendix 1A.

Table 4.1 highlights how different questions are related to the major sources and components of teachers' pedagogical knowledge in the conceptual framework.

The questionnaire does not include questions directly addressing the influence of teachers' experience as learners in informal educational environments on the development of their pedagogical knowledge. The main reason is that I believe this experience, compared with others, is a minor source for the subjects of this study who have at least college education experience and/or considerable teaching experience. However, for other kinds of teachers, such as prospective teachers (e.g., see Book *et al.*, 1983) or novice teachers without adequate education experience,[1] it might be an

---

[1] Note there are a considerable number of teachers of this type in many developing countries. It is also not rare that elementary teachers in those countries only have a very few years of educational experiences.

important source of their pedagogical knowledge. A second reason is that, unlike other questions, such questions about "informal educational environment" are difficult and not appropriate to be included in the questionnaire. Nonetheless, I did include questions relevant to this issue in the interview.

A pilot test of the questionnaire with three mathematics teachers selected from a school in the population, but not in the sample, showed that the questionnaire could be reasonably expected to be finished in about 20–25 minutes.

It needs to be pointed out that the results of the questionnaire can only provide part of a general picture of how teachers developed their pedagogical knowledge; particularly, because the lengths of both individual questions and the whole questionnaire should be reasonably short in order to get a good response rate to the questions and to the whole questionnaire, not all questions are appropriate to be asked in a questionnaire. This is a main reason I also designed research tools for classroom observation and interviews.

Classroom observation

If the questionnaire deals with what pedagogical knowledge teachers are concious of having, classroom observation is to identify, from the perspective of pedagogy, how teachers actually teach in classrooms and what kind of pedagogical knowledge teachers actually utilize in their teaching, so questions can be raised during the interviews about how those teachers developed relevant pedagogical knowledge. Classroom observation was used as an instrument for this study because, as mentioned earlier, teaching is a complex activity, and much of teachers' knowledge is "tacit" and "situated", and often can only be reflected and surfaced in actual classroom contexts (e.g., see Alexander *et al.*, 1991; Orton, 1993; Beijaard & Verloop, 1996). A further epistemological ground is that, as it has been widely recognized, people know more than they can tell because either they do not know what they know or they are not able to verbally express what they know; instead, their behavior or performance can often display what they really know (e.g., see Ryle, 1949; Ayer, 1956; Polanyi, 1966; Machlup, 1980).

As said previously, three teachers in each school with different teaching experiences were each observed for their actual classroom teaching twice. The classroom observation was designed to focus on the pedagogical aspects, namely, how teachers use the teaching resources and which ones they use (including textbooks, concrete materials, computers, and calculators), represent mathematical concepts or procedures to students, and employ teaching strategies and organize the classroom. To get this kind of information, I chose to observe normal lessons, that is, not specifically devoted to review or exams.

A guide for classroom observation can be found in Appendix 1B. All classroom observations were documented with audio recorders and field notes.

Interview

After classroom observation, structured interviews were conducted for the study. The interviewees included all the teachers who were observed, as well as the chairs of mathematics departments, a total of 12 individuals.

Three kinds of questions were prepared for the interviews with different interviewees.

The first kind, for the teachers, are questions about the lessons observed, focusing on why the teachers taught the lessons the way they did and how they acquired the knowledge they employed to plan the lessons. The questions were centered on PCrK, PCnK, and PIK, respectively. For example, concerning their PIK, all teachers were basically asked the question: "I noticed that you used the following teaching strategies (e.g., cooperative learning) in the classes I observed. How did you get to know these strategies?" Actual questions presented depended on the classroom context.

The second group of questions, for both the nine teachers and the three math chairs as teachers, are those not included in the questionnaire because of the reasons discussed earlier, but I think are helpful to know for the purposes of this study. For example, "It is said that teachers teach the way they were taught. Do you agree or not? Why? Can you identify one teacher or more when you were at school that significantly influenced the way you teach? (If yes) What are the influences?"

The last, for the three math chairs, are questions about inservice training and professional activities in their schools, which can help us to understand better the school environments within which teachers develop their pedagogical knowledge, and interpret teachers' responses to the questions in the questionnaire and in the interview. For example, "How do you organize professional activities in your department? What are the usual forms?"

The scripts for interviewing teachers and math chairs can be found in Appendices 1C and 1D. From the beginning, however, they were designed only to provide a general guidance, not a precise prescription for the actual interview, as an interview is an on-going interaction between the interviewer and interviewee, and how the questions are asked must depend on what knowledge the teacher demonstrates in the classrooms observed and other actual situations.

All interviews were audiotaped and transcribed for analysis. Each interview was scheduled to take about 45 minutes.

At this point, I want to stress that all instruments described above were not intended to detect how much knowledge teachers have, or how good teachers' knowledge is, which is beyond the purpose of this study; even though there are a few questions relevant to them (such as Questions 18a and 18b in the questionnaire, which ask teachers what textbooks they have used recently and how good they feel their knowledge of the textbook is), they were intended to provide relevant information that is needed for us to understand and analyze the teachers' responses, and/or to serve as stimulants for teachers to recall their experiences to answer other questions that are directly targeted at how teachers develop their pedagogical knowledge.

Pilot test of the instruments

The earlier versions of the instruments aforesaid were tested on a small scale to see how they could be improved and implemented.

1. Questionnaire. As mentioned earlier, three mathematics teachers including the math chair in one high school, which belongs to the population but not the sample of this study, were administered a nearly-final version of the questionnaire. After collecting the questionnaire, I had a discussion of about 30 minutes with all of them focusing on the practicability of the questionnaire.
2. Classroom observation. Three mathematics teachers' classroom teaching in a Chicago school was observed, and the three classrooms observations were videotaped.

3. Interview. Two teachers, one the male who answered an initial version of the questionnaire and the other a female whose classroom was observed, were interviewed. A draft version of the scripts for interviewing teachers was used for the interview. The interview with the female teacher was audiotaped.

These pilots were very helpful in the development of the instruments. The earlier versions of the instruments were further modified and refined using the information obtained from these pilot tests, though they revealed that the instruments were, overall, workable both conceptually and technically.

Data Collection

For statistical strictness, a TI-92 calculator was used to conduct the random sampling of schools from the population, with all the 25 schools being ranked from 1 to 25, based on their students' average scores in the IGAP mathematics test.

After the first three schools were chosen, an invitation letter was sent out to the chair of the mathematics department in each school, asking that the school participate in the study. However, only two schools responded positively, and the third found it difficult to do so as the chairman of the mathematics department of the school was leaving his job during the scheduled time for data collection (see below) and could not make the decision for his successor, so another school in the same stratum was again randomly selected using the same technique, which, upon my request, turned out willing to participate in the study. In order to keep their names anonymous, below I call the three schools participating in this study School 1A, School 1B, and School 1C, respectively.

Although there are slight differences between the three schools, broadly speaking all are typically good suburban public schools with a large predominantly white student population (ranging from 72% to 91%), advantageous teaching facilities, and good school environments, with the operational expenditures per pupil in the previous school year being between $10,000 and $12,000.

The numbers of students, teachers, and mathematics teachers in those schools are presented in Table 4.2.

Table 4.2 Numbers of students, teachers, and mathematics teachers in the three sample schools (1996–1997)

	School 1A	School 1B	School 1C	Total
Student population	3,124	2,906	2,308	8,338
No. of faculty members	233 (87%*)	260 (80%+)	180 (82%)	673
No. of mathematics teachers	31	24	22	77

*The percentage of those having at least one master's degree.

Table 4.3 Response rate of the questionnaire survey

	School 1A	School 1B	School 1C	Total
Number of mathematics teachers who returned the questionnaire	26	22	21	69
Response rate	83.9%	91.7%	95.5%	89.6%

After the school sample was decided, the questionnaire was distributed with the help of the math chairs to all mathematics teachers in these schools in late May 1997, and collected in mid-to-late June 1997. In total, the response rate was 89.6%, a very high level which suggests high dependability of the results according to common measures for a questionnaire response rate (e.g., see Borg, Gall, & Gall, 1993, p. 113; Engelhart, 1972, p. 96).[2] Table 4.3 shows the response rate in each school.

Based on teachers' responses to Questions 1–4 in the questionnaire, a profile of those 69 teachers is presented in Appendix 1E, which provides general information about those teachers' ages, genders, degrees earned, majors and minors of the degrees, and the lengths of their teaching experiences.

According to the data obtained from Question 4, a stratified random sample, discussed above, of nine teachers for classroom observation were selected. However, the initial sample included a teacher who was under the

---

[2]According to educational methodologists, the response rate of a questionnaire is related to a variety of factors including the questionnaire's length, question content and format, cover letter, survey time, and the surveyed. In this study, I believe the help from the math chairs was one crucial reason, amongst others, for the high response rate.

review for tenure. The teacher was then replaced by another teacher, again randomly selected. The decision was based on the math chair's suggestion to avoid the possible pressure that could be on the teacher under review if asked to be a volunteer for the study.

Then, the nine teachers were contacted to see if they could be volunteers for the classroom observation and interview. The purpose of the observation and interview was explained to each teacher. In particular, they were informed that I planned to observe their two normal classes (not for review or test), and hence required no special preparation for the classes to be observed. In addition, I emphasized that their names would not be released in the research report. All of the teachers agreed to participate in the study.

All of the nine teachers finally chosen were white. Each school had two females and one male participating. For both anonymity and convenience, I hereafter represent those teachers in School 1A as TA1 for the teacher with 0–5 years' teaching experience, TA2 with 6–15 years', and TA3 with 16 or more years', and those in School 1B, correspondingly, as TB1, TB2, and TB3, and those in School 1C as TC1, TC2, and TC3.

Table 4.4 contains information relevant to the selection of those teachers.

The actual classroom observation and interview were intensively conducted in September 1997. The classes observed are listed in Table 4.5.

Table 4.4 Selection of teachers for classroom observation

Teaching experience	0–5 years	6–15 years	16+ years	Average teaching experience (years)
School 1A (no. of math teachers)	7	9	10	13.3
Teacher selected (years of teaching mathematics)	*TA1* *(3 years)*	*TA2* *(9 years)*	*TA3* *(25 years)*	*(12.3)*
School 1B (no. of math teachers)	4	4	14	18.5
Teacher selected (years of teaching mathematics)	*TB1* *(2 years)*	*TB2* *(6 years)*	*TB3* *(27 years)*	*(11.7)*
School 1C (no. of math teachers)	6	4	11	16.5
Teacher selected (years of teaching mathematics)	*TC1* *(3 years)*	*TC2* *(14 years)*	*TC3* *(20 years)*	*(12.3)*
Total *(selected)*	17(3)	17(3)	35(3)	69(9); 15.9 *(12.1)*

Table 4.5 Classes observed for the study

School	Teachers	Classes observed
School 1A	TA1	Algebra C*; Geometry Honors
	TA2	Algebra C*; Precalculus Honors
	TA3	Calculus BC; Geometry Honors
School 1B	TB1	Geometry Preparatory*; Algebra IA
	TB2	Algebra II Preparatory*; Calculus AB
	TB3	Geometry Preparatory*; Geometry Acceleration
School 1C	TC1	Algebra; Precalculus/Discrete Mathematics
	TC2	Advanced Algebra; Precalculus/Statistics
	TC3	Geometry; Algebra IIIG*

Note: All classes marked with "*" consist of lower-learner students. Others consist of intermediate-level students or the highest-level students.

All classes were double audiorecorded to keep the original information as complete and secure as possible.

I intentionally observed two different classes with each teacher in the hope that he/she would display a wider range of pedagogical knowledge for different subjects and with a different cohort of students so that how he/she developed such knowledge could be raised in the interview. From the table, it can be seen that overall a variety of classes with various subjects and students were observed.

The interview with each of the nine teachers was held after his/her two classes were observed. Before the interview, the information relevant to the observed classes such as student background, the structures and characteristics of the textbooks used, and the teacher's lesson plans and teaching procedures were gathered and analyzed; more importantly, the recorded tapes were reviewed two or three times to familiarize the researcher (myself) with the classes observed so the concrete questions for the interview could be finalized. To have enough time to prepare for the interview, I arranged to interview every teacher one day after the second class was observed, except for two teachers, both of whom were interviewed on the same day as the second class was observed, and there was only a two-to-three-hour interval between the classroom observation and the interview; the time interval was exclusively used to analyze the second class observed and prepare for the forthcoming interview.

The interview with each math chair occurred on the last day of the field survey in the school. As one chair was currently not teaching classes because of a heavy administrative responsibility, only the other two chairs were asked the second type of questions (see earlier in this chapter), to which they answered from the perspective of being a mathematics teacher. The third type of questions, which are central for the interview with mathematics chairs, were raised for all three chairs.

All interviews were also double tape-recorded for the same reason as in classroom observation. The average time of all interviews was about 45 minutes, with the shortest about 35 minutes and the longest 75 minutes.

Data Processing and Analysis

The data collected from the questionnaire were first read and examined by this author. For organizational purposes, all questionnaires were numbered, with 1–26 for those from School 1A, 27–47 for School 1C, and 48–69 for School 1B.

The examination of the questionnaires proved that, in a few cases, some correctional treatment of the raw data was both necessary and practical. For example, for Question 4:

For how many years have you taught: Any subjects _____ years; Mathematics (including statistics, but excluding computer courses) _____ years.

Although most teachers filled in both blanks, some only chose to fill in the blank for teaching mathematics. Nonetheless, for those who did not fill in the first blank, I could conclude that the years of their teaching any subjects are equal to the years of their teaching mathematics, because for some of those teachers the fact is deducible from the answers to other questions. For instance, Teacher 29 answered that his age is between 20 to 30 (Question 2), holds a bachelor's degree (Question 3), just began an inservice master's program in mathematics education (Question 11) and received no other professional training experiences (Question 12), and most importantly, taught mathematics for one year, so it is very reasonable to believe that he is a new teacher with one year's teaching experience. Also, Teacher 68 filled "_" in the first blank but 8.5 years for the second, hence it is obvious that the "_" means the teacher did not teach any other subjects. The deduction was confirmed by some teachers who were later interviewed.

Another example is Question 13a:

In the last five years, about how many times did you attend professional activities of local, state, and national/regional organizations.

Local or State _____ times, National/Regional _____ times

Teacher 14 filled in the second blank with two to three times. I decided to count the answer as 2.5 times when averaging all teachers' answers to this question, as I believe this treatment is better than excluding it (two to three times is useful information and should be reflected in the general pattern).

Needless to say, those answers that cannot be adequately recovered (or are blank or unexpected) are excluded from the analysis. All of them, either corrected or simply excluded, are explicitly indicated in the findings reported in the following chapters.

After initial examination, all the data from the questionnaire were then stored, processed, and analyzed using a Macintosh computer, Microsoft Excel, and SAS.

As discussed earlier, the data collected from the classroom observation were mainly used to prepare questions for the following interview, as it only deals with what knowledge teachers have, not how they developed such knowledge. However, when analyzing the interview data, the tape records for the classroom observation were reviewed and utilized to help to understand teachers' answers in the interview.

The data obtained from the interview were transcribed verbatim via a professional transcription service firm and rechecked and then coded by this author.

The coding system for data processing was designed on the basis of the conceptual framework explicated in Chapter 3. Namely, teachers' pedagogical knowledge was categorized into PCrK, PCnK, and PIK. The sources of a teacher's pedagogical knowledge were first classified into his/her experience as a learner, preservice experience, and inservice experience, and then subcoded into more specific sources, including his/her experience as a student, preservice training, inservice training, (attending) organized professional activities, informal (daily) exchanges with colleagues, reading professional journals and books, and own teaching experience (practices) and reflection (see details in Chapter 3). In the initial reading of the questionnaires (mainly Questions 20 and 21 which are open-ended) and the transcriptions of the interviews, if teachers' responses could not be accounted for by the established coding system, another category was created and explained.

Two principles guided the data analysis.

First, both quantitative and qualitative methods were employed to analyze the data. The quantitative methods were used mainly on the data collected from the questionnaire to obtain some general patterns about how teachers developed their pedagogical knowledge. The qualitative methods were used mainly on the data collected from classroom observations and interviews to depict in-depth how individual teachers developed their particular pedagogical knowledge (e.g., when, where, and from whom did a teacher learn how to use, for instance, cooperative learning to teach mathematics?). I believe those two kinds of methods are confirmatory and complement each other.

Second, attention was also paid to the influence of teachers' background on the development of their pedagogical knowledge. To do that, the information of teachers' background, especially the length of their teaching experience, collected from the questionnaire was related to the results of teachers' evaluation about how they develop their pedagogical knowledge in the questionnaire. In particular, for example, within three different groups of teachers, one group with teachers who have taught mathematics for 16 or more years, one group between 6 and 15 years, and the other group 5 or less years, is there a significant difference how they view the importance of different sources (e.g., preservice training) to the development of their pedagogical knowledge?

The statistical techniques used to analyze the data are relatively simple and easily understood. Common statistical measures, such as mean, standard deviation, and percentage were used to describe the general pattern of how different sources contributed to the development of teachers' pedagogical knowledge. Log-linear models were utilized to depict more accurately if there existed significant differences between the contributions of different sources to teachers' pedagogical knowledge, and chi-square tests were employed to detect the influence of teachers' different backgrounds on the development of their pedagogical knowledge.

Strengths and Limitations of the Methodology

This study attacks an important issue in teacher education and teacher professional development. The strengths of the study are found by comparing it with other studies in this line reviewed in Chapter 2. Methodologically, it has three major distinctions from most of the previous work: 1. more specific — it directly addresses the question of how teachers develop their knowledge in the domain of pedagogy; 2. more systematic — it takes into account the whole life of teachers, including both their preservice and inservice experiences, not just a special period of teachers' lives and a special source of teachers' knowledge; and 3. more representative — the research subjects consist of a substantial number of teachers with all range of experiences, and selection of the sample is random in main stages.

Because of various factors such as time, expenses, and my own experience and knowledge, this study, like other studies, also has its limitations. Although the main conceptual framework might apply to teachers of all subjects, the actual subjects of the study are limited to high-school mathematics teachers, and therefore the results may not necessarily be applicable to teachers of other subjects or mathematics teachers of other grade levels. Moreover, the school sample is limited to the best public schools in a metropolitan area, so the results cannot be automatically extended to other kinds of schools such as urban or private schools, though I think the results might be applied with caution to the best public schools in other metropolitan areas.[3] From this point, similar studies which focus on different types of schools and teachers will be meaningful.

Finally, it has to be noted that all results of this study are based on the teachers' responses to the questionnaire and interview, which eventually come from their recall of their own experiences. Although stimulants and incentives were carefully designed in both the questionnaire and interview to solicit teachers' responses as accurately as possible, the data from them are still only an indicator of how teachers developed their pedagogical knowledge. Teachers, like other adults, may either remember or forget how they learned a particular piece of knowledge, or sometimes, even if they

---

[3] I provided earlier some general background information of the schools and teachers participating in this study so it can be potentially used to compare with other schools.

remember or forget, they can be either willing or unwilling to tell the truth. When doing survey research, one can only try to reduce as much as possible, but can never totally remove, these problems. Nonetheless, I believe teachers' own voices are an irreplaceable indicator of how they developed their own knowledge, and must be heard and studied.

Summary of the Methodology

The central question of this study is, how do teachers develop their pedagogical knowledge? To address this question, I take all the 77 mathematics teachers in three high-performing high schools, a stratified random sample from the 25 top-level high schools in terms of their students' average scores in the 1996 IGAP mathematics test in the metropolitan Chicago area, as the subjects of this study.

Three instruments were designed and employed in this study to collect the data.

The first was a questionnaire which consists of 22 questions and is based on the conceptual framework described in Chapter 3; the questionnaire was administered to all 77 mathematics teachers in the three schools and there was a response rate of 89.6%.

The second was classroom observation, designed to identify what kind of pedagogical knowledge teachers actually demonstrate and utilize in their teaching; the classroom observation was applied to nine teachers with three teachers from each school: one randomly selected from teachers with 0–5 years' teaching experience, one from those with 6–15 years' teaching experience, and the other 16 or more years' teaching experience. Two normal classes of each teacher were observed.

The third was an interview which was applied to the nine teachers observed and the three math chairs in these schools. The interviews with teachers mainly focus on how they learned the specific pedagogical knowledge demonstrated in the classes observed; while the interviews with math chairs focus on questions about the school professional environments for teachers to develop their pedagogical knowledge.

The raw data collected from the questionnaire were examined and the data from the interview were transcribed, and then, according to the conceptual framework discussed in Chapter 3, all the data were coded for data processing and analysis. Quantitative methods were mainly used on the data collected from the questionnaire to obtain some general patterns about how teachers developed their pedagogical knowledge. In contrast, qualitative methods were mainly used on the data collected from classroom observations and interviews to depict in-depth how certain teachers developed their particular pedagogical knowledge.

In addition, attention was also paid to the influence of teachers' background on the development of their pedagogical knowledge during the process of data analysis.

Chapter 5

Findings of the Chicago Study (I): Pedagogical Curricular Knowledge

The main findings of the Chicago study are reported in Chapters 5–8. Following the conceptual framework of the study discussed in Chapter 3, Chapters 5, 6, and 7 are devoted to the findings of how different sources contribute to the development of teachers' pedagogical curricular knowledge (PCrK), pedagogical content knowledge, and pedagogical instructional knowledge, respectively. Chapter 8 does not particularly separate teachers' pedagogical knowledge into the three different components and attends to some other issues, which are not discussed in Chapters 5–7, but are related to the theme of this study and helpful for us to understand the core findings reported in the previous three chapters.

In all four chapters, the findings are based on the data collected from the questionnaire survey and interviews. The data collected from the classroom observations are used to help explain and understand the findings as appropriate.

The focus of this chapter is on how different sources contribute to the development of teachers' PCrK. As explained earlier, teachers' PCrK is about instructional materials and teaching resources. In this study, teachers' pedagogical knowledge is investigated through their knowledge of teaching materials (primarily textbooks), of technology (primarily calculator and computer/software), and of other teaching resources (primarily concrete materials).

Knowledge of Teaching Materials

The core teaching materials for most mathematics teachers are textbooks, but textbooks, in a broader sense, also include reference books, problem booklets, and other materials. In fact, along with students' editions and teachers' editions of textbooks, many textbook developers and publishers also produce a variety of teaching materials centered on the main textbooks. For example, UCSMP's (University of Chicago School Mathematics Project) secondary textbooks, published by Scott Foresman–Addison Wesley, are accompanied by a series of teachers' reference materials, including *Lesson Masters*, *Teaching Aid Masters*, *Assessment Sourcebook*, *Technology Sourcebook*, *Answer Masters*, and so forth.[1] It seems appropriate to use "main textbooks" to refer to the core teaching materials and "peripheral textbooks" to the other teaching materials. Below I will use "textbooks" to generally refer to teaching materials.

The influence of textbooks on teachers' teaching practice has received increasing attention from researchers (e.g., see Krammer, 1985; Graybeal, 1988; Stodolsky, 1989; Sosniak & Stodolsky, 1993; Fan & Kaeley, 1998). Although there exists disagreement about the magnitude of the influence of textbooks, researchers generally agree that teachers rely heavily on textbooks in their day-to-day teaching, and they decide what to teach, how to teach it, and what sorts of exercises to assign to their students largely based on the textbooks they use (Robitaille & Travers, 1992). It seems to me, as textbooks themselves convey various pedagogical information and orientations to teachers, how much the textbooks teachers are using affect their teaching depends on how much teachers actually know about the textbooks, which is further related to how teachers get to know the textbooks, a theme of this study.

Question 18 of the questionnaire is about teachers' knowledge of textbooks. Part a of the question asks each teacher to fill in the name of a course which he/she taught in the most recent period, the title of the textbook for the course, and the number of years using the textbooks. Table 5.1 summarizes the relevant information provided by 67 of the 69 teachers who answered the questionnaire (two teachers did not provide information).

---

[1] A more detailed description of UCSMP teaching materials can be found on-line at http://ucsmp.uchicago.edu/newsletters.

Table 5.1 Courses taught by teachers

Algebra	Geometry	Advanced algebra	Precalculus (including statistics, trigonometry, and discrete mathematics)	Calculus
10	22	15	18	2

Note: n = 67. Courses combined in the cells include different levels, such as preparatory (for lower learners), acceleration (for regular learners), and honor class levels.

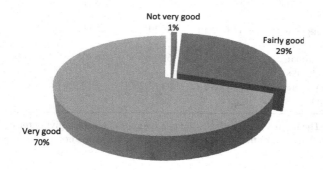

Figure 5.1 Knowledge of textbooks being used

Note: n = 68, as one teacher did not provide the information.

There were 29 different textbooks used by 66 teachers (three others did not provide the information). For 65 teachers who gave the exact numbers of years of using their textbooks, the length ranged from one year (15 teachers) to over 10 years (2 teachers), with a median of three years and an average of 3.2 years (standard deviation: 2.4 years).

Part b is about how teachers felt about their knowledge of the textbooks listed in Part a. The result is displayed in Figure 5.1.

Parts a and b provide a stimulant for teachers to answer, as well as a context for us to look at teachers' answers to Part c, which is our focus.

Part c asks teachers to evaluate the contributions of various sources to the development of their knowledge of the textbooks mentioned in Parts a and b. Table 5.2 depicts the distributions of teachers' responses.

Two initial results can be found from the above table. First, there are a variety of sources from which teachers can develop their knowledge of textbooks, and for different teachers, their main sources for developing their knowledge of textbooks could be very different. Second, in terms

Table 5.2 Distributions of the numbers of teachers giving different evaluations about the contribution of various sources to the development of their knowledge of textbooks

Sources	Degree of the contribution			
	Very much	Somewhat	Little	No contribution
Experience as student	4 (5.9%)	10 (14.7%)	13 (19.1%)	41 (60.3%)
Preservice training	1 (1.5%)	7 (10.3%)	14 (20.6%)	46 (67.6%)
Inservice training	13 (19.1%)	14 (20.6%)	11 (16.2%)	30 (44.1%)
Organized professional activities	5 (7.4%)	16 (23.5%)	13 (19.1%)	34 (50.0%)
Informal exchanges with colleagues	36 (52.9%)	20 (29.4%)	10 (14.7%)	2 (2.9%)
Reading professional journals and books	2 (2.9%)	6 (8.8%)	16 (23.5%)	44 (64.7%)
Own teaching practices and reflection	57 (83.8%)	9 (13.2%)	2 (2.9%)	0 (0%)

Note: n = 68. The figures in parentheses are percentages of teachers giving the corresponding evaluation. The sum of the percentages in each row might be not exactly 100% due to rounding.

of the combined percentages of teachers' choosing positive evaluation of "very much" and "somewhat", overall teachers' "own teaching practices and reflection" (97.0%) and "informal exchanges with colleagues" (82.3%) are the two most important ways in which those teachers gained their knowledge of textbooks, while "reading professional journals and books" (11.8%) and "preservice training" (11.8%) are the least important sources. The other three have moderate influences, with the percentages of the positive evaluation for "inservice training" being 39.7%, "organized professional activities" being 30.9%, and "experience as student" being 20.6%.

Figure 5.2 presents a general comparison of the contribution of different sources to teachers' knowledge of textbooks, which is based on their average evaluation in Part c.

According to teachers' average evaluation shown in the figure, we can see that the order of the sources for teachers to develop their knowledge of textbooks is, from the most important to the least important, teachers' "own teaching practices and reflections" (3.81), "informal exchanges with colleagues" (3.32), "inservice training" (2.15), "organized professional activities" (1.88), "experience as student" (1.66), "reading professional

Findings of the Chicago Study (I): Pedagogical Curricular Knowledge 79

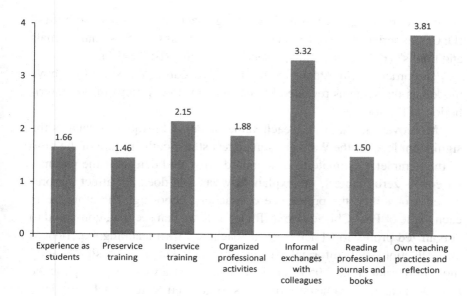

Figure 5.2 Comparison of the contribution of different sources to teachers' knowledge of textbooks

Note: By the ordinal scale in the figure, 4 = very much, 3 = somewhat, 2 = little, and 1 = no contribution.

journals and books" (1.50), and "preservice training" (1.46), consistent with the results obtained from Table 5.2.

Based on the above preliminary analysis of the data, log-linear regression models were further employed to analyze the data. Table 1F.1 in Appendix F displays the main result of "PROC LOGISTIC" procedure using SAS, which gives parameter estimates (maximum likelihood estimates) and statistical tests for the estimates.[2]

According to the parameter estimates shown in Table 1F.1, Appendix F, the order of importance of the sources to the development of teachers' knowledge of textbooks is teachers' "own teaching experiences and reflection" (G: −4.0523), "informal exchanges with colleagues" (E: −2.5298),

---

[2] Log-linear regression models have been increasingly used by researchers to analyze categorical data since the 1970s. Some texts that introduce log-linear models include Agresti (1990), Everitt (1992), Ishii-Kuntz (1994), Sobel (1995), Agresti (1996), Christensen (1997), and Long (1997).

"inservice training" (C: −0.4284), "organized professional activities" (D: 0), "experience as student" (A: 0.4551), "reading professional journals and books" (F: 0.7346), and "preservice training" (B: 0.8491).

The order of importance of the different sources based on the above models is the same as revealed by teachers' average ratings of the contribution of the sources.

Moreover, in Table 1F.1 each value of Pr > Chi-Square indicates the significant level of the Wald chi-square test statistic, the square of the ratio of the parameter estimate to its standard error, and detects if the parameter equals zero, namely, the explanatory variable does not affect the predicted probability, the preference of teachers' choosing. Therefore, from each value of Pr > Chi-Square in Table 1F.1, we can see that, compared to "organized professional activities",[3] teachers' "own teaching experiences and reflection" and "informal exchanges with colleagues" are significantly more important at the 0.05 level, their "inservice training" and "experience as student" have the same importance as "organized professional activities", and "reading professional journals and books" and "preservice training" are significantly less important.

Now, let me address how the length of teachers' teaching experience is related to their evaluation of the contribution of different sources to their knowledge of textbooks. As mentioned previously, the teachers were classified into three groups: TG1 consisting of teachers with 0–5 years' teaching experience, TG2 with 6–15 years', and TG3 with 16 or more years'. Based on that, I use a descriptive statistic, namely the average scores of teachers' evaluations, to measure their overall evaluations in the three groups and employ chi-square tests to see if there is a statistically significant difference among the groups.

Figure 5.3 presents the average evaluation of the teachers in those three groups of the contribution of different sources to their knowledge of textbooks; while Table 5.3 displays the actual distributions of the numbers of the three groups of teachers giving different evaluations of the contribution of each source to their knowledge of textbooks, as well as the results of the corresponding chi-square test for each source.

---

[3]Note: we set $D = 0$ in the models.

Findings of the Chicago Study (I): Pedagogical Curricular Knowledge 81

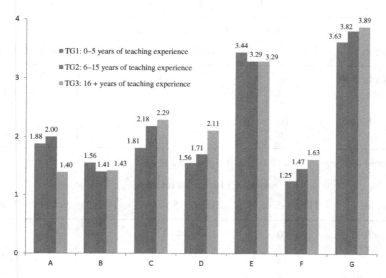

Figure 5.3 Comparison of the three groups of teachers' average evaluation of the contribution of different sources to their knowledge of textbooks

Note: 1. For sources, A = Experience as student, B = Preservice training, C = Inservice training, D = Organized professional activities, E = Informal exchanges with colleagues, F = Reading professional journals and books, G = Own teaching experience and reflection.

2. The evaluations are shown by the ordinal scale in the figure, 4 = very much, 3 = somewhat, 2 = little, and 1 = no contribution.

Table 5.3 Distributions of the numbers of teachers among the three groups giving different evaluations of the contribution of each source to their knowledge of textbooks

	Source A: Experience as student			
Evaluation	TG1(0–5 years)	TG2(6–15 years)	TG3(16+ years)	Total
Very much	1	2	1	4
Somewhat	4	4	2	10
Little	3	3	7	13
No contribution	8	8	25	41
Total	16	17	35	68

Chi-square test: $\chi^2 = 7.0547$ df $= 6$ p $= 0.3158$

(*Continued*)

Table 5.3 (*Continued*)

	Source B: Preservice training			
Evaluation	TG1(0–5 years)	TG2(6–15 years)	TG3(16+ years)	Total
Very much	0	0	1	1
Somewhat	3	2	2	7
Little	3	3	8	14
No contribution	10	12	24	46
Total	16	17	35	68

Chi-square test*: $\chi^2 = 1.2410$ df = 4 p = 0.8713

	Source C: Inservice training			
Evaluation	TG1(0–5 years)	TG2(6–15 years)	TG3(16+ years)	Total
Very much	0	4	9	13
Somewhat	5	3	6	14
Little	3	2	6	11
No contribution	8	8	14	30
Total	16	17	35	68

Chi-square test: $\chi^2 = 5.7688$ df = 6 p = 0.4496

	Source D: Organized professional activities			
Evaluation	TG1(0–5 years)	TG2(6–15 years)	TG3(16+ years)	Total
Very much	0	1	4	5
Somewhat	3	3	10	16
Little	3	3	7	13
No contribution	10	10	14	34
Total	16	17	35	68

Chi-square test: $\chi^2 = 4.3000$ df = 6 p = 0.6362

	Source E: Informal exchanges with colleagues			
Evaluation	TG1(0–5 years)	TG2(6–15 years)	TG3(16+ years)	Total
Very much	9	10	17	36
Somewhat	5	3	12	20
Little	2	3	5	10
No contribution	0	1	1	2
Total	16	17	35	68

Chi-square test: $\chi^2 = 2.5001$ df = 6 p = 0.8685

(*Continued*)

Table 5.3 (*Continued*)

Source F: Reading professional journals and books

Evaluation	TG1(0–5 years)	TG2(6–15 years)	TG3(16+ years)	Total
Very much	0	0	2	2
Somewhat	0	3	3	6
Little	4	2	10	16
No contribution	12	12	20	44
Total	16	17	35	68

Chi-square test: $\chi^2 = 6.8552$ df = 6   p = 0.3344

Source G: Own teaching practices and reflection

Evaluation	TG1(0–5 years)	TG2(6–15 years)	TG3(16+ years)	Total
Very much	12	14	31	57
Somewhat	2	3	4	9
Little	2	0	0	2
No contribution	0	0	0	0
Total	16	17	35	68

Chi-square test**: $\chi^2 = 7.0900$ df = 4   p = 0.1312

* Rows 1 and 2 are combined for the chi-square test, for the original 4 × 3 table produces three expected cell frequencies less than 1. The combined 3 × 3 table nicely meets the common assumption of using the chi-square test.[4]
**Because here all the cells in the last row are 0, the chi-square test cannot be directly applied. The results were obtained by applying the test to the rest of the table, excluding the last row.

From Figure 5.3, we can see that overall the average evaluations of those three groups of teachers for each of the sources are rather close. This is verified by the p-values of the chi-square test shown in Table 5.3, for there is no significant difference among the distributions with those three groups for each source. In other words, the length of teachers' teaching experience did not significantly affect how they think about the importance (contribution) of different sources in developing their knowledge of textbooks.

---

[4] For example, see Moore & McCabe (1993) and Peers (1996).

Compared with the questionnaire data, the interviews with the nine teachers revealed more specific and contextualized evidence of how different sources contributed to the teachers' knowledge of textbooks.

In the interviews, all the nine teachers were first asked how they developed their knowledge of the textbooks that were being used in the classes observed. When it was unclear whether or not they had described all the sources for them to develop the knowledge, they were asked if they had other sources to make sure they could make the list of sources as complete as possible. The same interview methodology applied to teachers' knowledge of other components (such as of technology, of concrete materials, and so on), which are discussed later in this chapter and in the following chapters.

Below is a description, source by source, of how teachers answered the interview questions. Occasionally, some quotes, especially those included in parentheses, are not particular to the source being discussed; instead they are stated here to give background.

Experience as student. Only one teacher (TB1) described experience as a school student as a source that contributed to his knowledge of the textbooks used in the observed classes. This teacher had two years' teaching experience. The following is an excerpt[5] from the interview record between the interviewer (myself) and that teacher. The textbook discussed is *Geometry: For Enjoyment and Challenge*, authored by Rhoad, Milauskas, and Whipple (1991).

> Interviewer: I noticed that you chose different examples and in-class exercises from the textbook; also the assignments were different from the textbook. I think you know pretty much of the textbook. So my question is, how did you get your knowledge of the textbook, like the structure, the lessons, the arrangement, exercises, and questions of the textbook?
>
> TB1: Well, that's actually the textbook that I used when I was in high school geometry. So I was very familiar with that one.... I was taught out of that one in 1984, fourteen years ago.
>
> Interviewer: So maybe it's a new edition.
>
> TB1: Yes, a newer edition. It's just a different edition, but it's pretty much word for word the same.

---

[5] All excerpts in this book are condensed from the original records for the purpose of saving space and keeping the text focused on the research questions.

It was the first time that this teacher taught the textbook, and the other source he mentioned is "reading the text by myself".

It seems clear that whether a teacher's experience as a student contributed to his/her knowledge of a textbook depends on how the textbook he/she is teaching is related to the one he/she was taught. If they are the same,[6] the contribution is big; if they are closely related, such as an early edition and a later edition like the case above, the contribution is also considerable. However, if they are different, the contribution is little, and teachers have to resort to other sources to develop their knowledge of the textbooks they will teach. Because in the US, there are so many textbooks available and textbooks change relatively quickly, except for some special cases, it is unlikely that teachers, especially senior teachers, teach the textbooks they were taught. This might explain why most teachers interviewed did not include their experiences as students as a source of their knowledge of the textbooks they taught.

Preservice training. Only one teacher, TA1, answered that her preservice experience helped her to develop her knowledge of the textbooks. While stressing the importance of informal exchanges with her colleagues (see later in this section), she explained how her preservice experience had been helpful for her to get knowledge of different textbooks:

> TA1: I did do a little bit [with textbooks] in college. We got together with a group of people. We had a working group of like four students. And we had to come up with an evaluation sheet for math textbooks and decide what we liked and what we didn't like. Things that we thought were good and things we didn't think good. And we had to write them. Once we had our evaluation, we went in one day, and there was just a stack of books around the room, and we had to go through them and do evaluations for all of them. That helped me look for things like that is a good idea that I'd like to incorporate into my classroom, and that is something that the textbook is talking about and it probably isn't that valuable. So this kind of helps me out.

No other teacher listed preservice experiences as a source that contributed to his/her knowledge of the textbooks he/she was teaching.

---

[6]This is especially the case in countries like China, where I was taught and also taught for many years. There was once only one series of textbooks for the whole country and revision of the textbooks was slow.

Inservice training. According to the questionnaire data, only a small number of teachers received inservice training focusing on textbooks and other teaching resources, with the percentage in the last five years being 21.7% (also see Chapter 8, p. XX). This is largely consistent with the interview data, which revealed that most teachers did not receive inservice training for the textbooks they were teaching. Only one teacher, TB3, pointed out that attending professional training had been useful to enhance her knowledge of the textbooks.

> [After TB3 identified that her own teaching experience and reflection had been the major source for her to develop her knowledge of textbooks and teaching materials]
>
> Interviewer: Are there any other sources?
>
> TB3: Well, my other sources would be sometimes (colleagues). Or in the summertime, I go to workshops. This summer I went to Illinois State University for two classes. Sometimes I go down to Champaign. They have a little bit of everything for textbooks or teaching strategies. Sometimes it's on algebra. Sometimes geometry. Sometimes calculators.

Organized professional activities. One teacher, TB2, who was teaching a reform textbook, *Calculus: Concepts and Applications* (Foerster, 1998), said that she attended two or three different conferences in the past year and had spent a week at a program that summer on the calculus reform, and that experience was useful for her in developing her knowledge of the reform textbook she was teaching. She was the only one recognizing organized professional activities as a source for developing knowledge of textbooks.

The reason that not many teachers listed attending organized professional activities as a source of their knowledge of textbooks seems to be that not many of those kinds of activities are available which focus on textbooks.

Informal exchanges with colleagues. Six of the nine teachers pointed out that they got their knowledge of the textbooks they were teaching from their colleagues.

The following is a discussion between the interviewer and teacher TA1 about both her specific knowledge of the textbook's arrangement of a special topic: velocity and speed, and her general knowledge of the textbook, *Algebra 1: An Integrated Approach* (Larson, Kanold, & Stiff, 1995a, pp. 62–67).

> Interviewer: In the algebra class I observed, the textbook has "Example 6: Finding velocity and speed". You did not include it in the class. This is because you did not have enough time?
>
> TA1: A little bit. More because later on in algebra [class,] speed and velocity equations will come up again....
>
> Interviewer: Why do you know it will come up again in the textbook?
>
> TA1: Because I taught it before. I had experience.
>
> Interviewer: But if it was your first time you taught the textbook?
>
> TA1: Then it was my colleagues helping me out by telling me: don't worry about this right now; we'll get to it in Chapter Four when it comes up again.
> ...
>
> Interviewer: How did you develop your general knowledge of the textbooks?
>
> TA1: (I did do a little bit in college...[7]). But again too, I'd have to say that a lot of it comes from my colleagues, who have been through it already and can tell you things. Probably the first time looking through that textbook and seeing velocity, I wouldn't have known that it would come up again later. But they are there to tell you that this point we can rely on later.

TC1's precalculus class and algebra class, both of which he was teaching for the first year, were observed. He mentioned several times how he acquired knowledge of the textbooks from his colleagues, which are UCSMP *Precalculus and Discrete Mathematics* (Peressini et al., 1992) for the precalculus class, and UCSMP *Algebra* (McConnell et al., 1996) for the algebra class.

> Interviewer: Who sets the purposes of the course [precalculus]?
>
> TC1: The school. The math department.
>
> Interviewer: How did you know those purposes?
>
> TC1: OK. The textbook has very basic objectives for each lesson. We decide which lessons are going to come out of the book that we're going to teach.... When I do it, I went to a previous teacher. He's taught it for several years. And he said, OK. Here's the order we go through as far as chapters and the material we're trying to cover. The book is very good about having good objective in there. But sometimes I feel like the book skips algebra concepts that students need or doesn't hit them hard enough. It just kind of brushes them.

---

[7] See the earlier section about preservice training.

> Interviewer: How did you know that?
>
> TC1: OK. I've been told by other teachers. But I've also noticed the lack of algebra skills in these students.
>
> ...
>
> Interviewer: How did you develop your general knowledge of the textbooks?
>
> TC1: First I read them over the summer by myself... Now as you go through there, you look at concepts and you say, OK, is that a key concept or not? For that kind of knowledge, I rely mainly on some of [the] other teachers now, because I don't have a great feel for that yet.
>
> Interviewer: So you had exchanges with your colleagues about the textbooks?
>
> TC1: Yes. I talked about — like I said, I found out last year I was going to teach this class, so I went to the people who were teaching it last year, and talked to them about, what are the key concepts? What chapters are we going to hit?

Four other teachers, not quoted above, also answered that they learned knowledge of the textbooks from their exchanges with their colleagues.

Reading professional journals and books. No teacher interviewed explicitly described "reading professional journals and books" as a source of their knowledge of the textbooks they were using. Comparing this to the result of the questionnaire survey, we can be quite sure that it is by no means a major source. I think there might be two reasons. One is that teachers did not read professional journals and books very often (see details in Chapter 8); the other is that the professional journals and books teachers have access to, such as *Mathematics Teachers*, are usually devoted to mathematics content and general instructional issues, not to textbooks.

Own teaching experience and reflection. Except for the two teachers, TB1 and TC1, who were teaching the textbooks for the classes I observed for the first time, all of the other teachers interviewed expressed that their own teaching experience and reflection had contributed to their knowledge of the textbooks.

Below are excerpts from the interviews with TA3 and TB3, who gave more detailed explanations than other interviewees about the issue.

TA3 has taught mathematics for more than 20 years. One of her classes I observed was calculus; the textbook was *Calculus with Analytic Geometry* (Larson, Hostetler, & Edwards, 1994). The other was a geometry honors class; the textbook was *Geometry: For Enjoyment and Challenge* (Rhoad,

Milauskas, & Whipple, 1991). The conversation began with the calculus class.

> Interviewer: How did you get the general knowledge of the textbooks?
>
> TA3: I've taught calculus out of many textbooks. This is probably about the fifth book I've taught calculus from. So I've seen a lot of textbooks.
>
> Interviewer: How about geometry?
>
> TA3: Actually, I taught geometry 18 years ago. I didn't teach geometry for a long time. I just went back to teaching geometry three years ago. I would say that my view of geometry changed because I teach the classes that come after geometry.
>
> Interviewer: So you mean you learned that from your teaching experience?
>
> TA3: Yes.
>
> Interviewer: Did you get any of that kind of knowledge of textbooks from other sources?
>
> TA3: No. No. I don't read textbooks very carefully.
>
> Interviewer: But you do know what's good and what's not very good, what's the limitation of the textbooks. Is that right?
>
> TA3: Yes.
>
> Interviewer: So you know that from your experience?
>
> TA3: Experience. Yes. I have a lot of experience. I've been teaching for 22 years, so I have a lot of experience.

TB3 has 27 years of teaching experience. The classes observed were both geometry, but one for lower-level students and the other for regular students. Both classes were using the same textbook: *Geometry: For Enjoyment and Challenge* (Rhoad, Milauskas, & Whipple, 1991). The teacher explained how she developed her knowledge of textbooks from her experience.

> Interviewer: In your classes I observed, I noticed you used different examples and questions from the textbooks, but you also assigned students some questions in the textbooks. You must have certain knowledge of the textbooks. My question is, how did you develop that kind of knowledge?
>
> TB3: I have taught that course more than one time. The textbook is not new to me. So after I know what the textbook is like, then I know which problems the students had trouble with last year.
>
> Interviewer: That's from your own experience?
>
> TB3: That was from my own experience from having taught it [the] previous year. When I've taught it the previous year, then I make a note in the textbook, or actually on my assignment sheet, as you've probably noticed. ...When I know a problem has given my class trouble, then I circle the problem. And the next

year, I make sure I pay special attention to the really hard ones that everyone was frustrated with.

Interviewer: How many years have you taught out of this textbook?

TB3: Maybe three years.

Interviewer: So it is different from the first year you taught?

TB3: Usually. That's why I think it's extremely poor to have a teacher teach one year and never teach it again. Because everything they learn and they think will be not used again. Like this is a bad question in the book; it was silly. You know, you don't sometimes know how poor a question is and how confused the students are until you've presented it. Now in most textbooks, they suggest what to give. I don't know if you've seen the teachers' editions.

Interviewer: No. I didn't see it.

TB3: They give you suggestions with problems. The first year I usually follow their suggestions. However after I follow their suggestions, a lot of times I don't like the quality of the problem, and that's when I'd start giving them extra sheets. Now luckily, because I'm an older teacher, even when I get a new book, I have old papers from previous books. I didn't throw anything away. So let's say next year, people don't want this book, and we get a new book. Geometry is still geometry. Just because it's Chapter 4 in this book, it may be Chapter 2 in another book. Although the problems won't match, the chapters will still match. Your experience still helps.

Interviewer: So can I say your own teaching experience and reflection about how to use the textbook is a major source for you today?

TB3: Absolutely.

Interviewer: Are there any other sources?

TB3: Well, my other sources would be sometimes [informal exchanges with colleagues and attending summertime workshops].

Other sources. In addition to what is discussed above, the interviews also revealed that there are two other kinds of sources, which were not originally highlighted in the conceptual framework established in Chapter 3.

The first source is "reading textbooks". Needless to say, a teacher can get knowledge of textbooks from reading them, for example, one can know roughly the structure of a chapter in a textbook from reading the chapter. In the interviews, not all teachers particularly described "reading textbooks" as a source of their knowledge of the textbooks, and two teachers even explicitly pointed out that they usually did not read the textbooks very much because they had taught the textbooks previously. Nonetheless, five teachers stressed that when they taught a course for the first time, reading the textbook was very important for them to get to know the textbook.

Look at how some of them explained it below.

> Interviewer: How did you get all of that kind of knowledge of the textbook [calculus]?
>
> TB2: Once again, what I did was, this summer I spent three months with the textbook. With the calculus book I went through each section and I did [a] handful of problems for each section, and kind of got a feeling of what it was going to be about it. So, that's the first thing I did. I went through the textbook as though I was a student and did the problems myself.
>
> ...
>
> Interviewer: How did you develop your knowledge of the teaching resources [a UCSMP textbook]?
>
> TC2: Well, I read it in the teachers' edition. My business experience [see below on this page] helped me because I helped that textbook, but that was secondary. The thing was, I can't remember absolutely everything, and so I read. Bottom line is I read the teachers' edition.
>
> ...
>
> TB3: ... But if it were a brand new book, then in the summertime, I would sit down and I would read the beginning of every chapter. In [the] teachers' edition, they say how long it should take for superiors. And then, I map out my lessons in the summertime. That doesn't mean that they'll stay that way, but it gives me a rough idea.

It is clear that "reading textbooks" is especially important for teachers who teach the textbooks for the first time, and that it remains helpful every time they teach. Because in some sense it is teachers' current experience, I believe it is appropriate to treat it as a kind of teachers' "own teaching experience and reflection" under the context of this study.

The second source in particular is found with TC2, who worked as an editor of secondary mathematics textbooks at a publishing company. Therefore, she described her "business experience" as an important source of her knowledge of the textbooks. However, it should be pointed out that not only is the number of teachers who had worked in other careers before getting teaching jobs very small, but also understandably this kind of "business experience" is likely unrelated to school textbooks. So TC2's case is very unusual. Nonetheless, it reminds us that there are indeed various sources for different teachers to develop their knowledge of textbooks.

Table 5.4 summarizes the data from the interviews about the question.

To conclude this section, from both the questionnaire and the interview data, it is clear that there are various sources by which teachers can

Table 5.4 Sources of teachers' knowledge of textbooks (from interview data)

	A	B	C	D	E	F	G	Other sources
TA1		yes			yes		yes	
TB1	yes							Reading text
TC1					yes			Reading text
TA2					yes		yes	
TB2				yes	yes		yes	Reading text
TC2					yes		yes	Reading text; Business experience
TA3							yes	
TB3			yes		yes		yes	Reading text
TC3							yes	
Total	1	1	1	1	6	0	7	

Note: For sources, A = Experience as student, B = Preservice training, C = Inservice training, D = Organized professional activities, E = Informal exchanges with colleagues, F = Reading professional journals and books, G = Own teaching experience and reflection.

develop their knowledge of textbooks. However, overall, the most important sources are teachers' own teaching experience and reflection and their daily exchanges with their colleagues; teachers' attending organized professional activities, inservice training, and experience as school students are the secondarily important sources; and their reading professional journals and books, and preservice training are the least important ones. Statistically, the contributions of the three kinds of sources to teachers' knowledge of textbooks are significantly different. Moreover, teachers' teaching experience did not play a significant role in their viewing of the contributions of those sources to their knowledge of textbooks.

Knowledge of Technology

In this study, I follow the NCTM *Standards*, in which "technology" is defined as one of the instructional resources teachers need to use in their classroom teaching, which most commonly and importantly includes calculators and computers (computer software), although some researchers have also nicely defined "technology" as "electronic software and hardware" (e.g., see Heid, 1997). Below, I shall mainly focus on calculators and computers when mentioning technology. Undoubtedly, it is this kind of computing technology that has greatly influenced what to teach and how to teach in school mathematics (for a relatively comprehensive review of that issue, see Kaput, 1992).

In a sense, teachers' knowledge of technology can be separated into two components: knowledge of technology itself and knowledge of how to use technology in their instruction. Nonetheless, when I report the findings of this study hereafter, I shall generally include both components of teachers' knowledge of technology as their PCrK. The main reason is that although anyone (not necessarily a teacher) can have the first component, which can be independent of the second component, the second component cannot be independent of the first. In other words, it is unlikely that a teacher knows how to use technology (e.g., a graphing calculator) to teach mathematics without first knowing the technology itself. To teachers, those two components are closely related and hard to separate. As a matter of fact, one teacher interviewed in this study clearly pointed out that in seminars she attended about technology, they "not only teach you how to use calculators, but also give you opportunities to discuss what is appropriate in the classroom".

Part a of Question 16 in the questionnaire asks how often teachers used computers and calculators during the school year. Table 5.5 describes the results from the sample of teachers. The results suggest that all the teachers have at least sometimes used calculators and slightly more than 50% of teachers have used computers in classrooms. The difference of the percentage of teachers using calculators and that using computers is considerable.

Table 5.5 How often teachers used calculators and computers in their mathematics classes

	Never	Rarely	Sometimes	Most of the time	Always
Calculator (n = 69)	0 (0.0%)	0 (0.0%)	5 (7.2%)	27 (39.1%)	37 (53.6%)
Computer (n = 68)	3 (4.4%)	29 (42.6%)	29 (42.6%)	5 (7.4%)	2 (2.9%)

Table 5.6 Distributions of the numbers of teachers giving different evaluation about the contribution of various sources to the development of their knowledge of how to use technology for teaching mathematics

Sources	Degree of the contribution			
	Very much	Somewhat	Little	No contribution
Experience as student	2 (3.0%)	12 (17.9%)	18 (26.9%)	35 (52.2%)
Preservice training	1 (1.5%)	13 (19.4%)	15 (22.4%)	38 (56.7%)
Inservice training	35 (52.2%)	24 (35.8%)	5 (7.5%)	3 (4.5%)
Organized professional activities	20 (29.9%)	33 (49.3%)	10 (14.9%)	4 (6.0%)
Informal exchanges with colleagues	46 (68.7%)	17 (25.4%)	4 (6.0%)	0 (0%)
Reading professional journals and books	2 (3.0%)	21 (31.3%)	26 (38.8%)	18 (26.9%)
Own teaching practices and reflection	49 (73.1%)	17 (25.4%)	1 (1.5%)	0 (0%)

Note: n = 67. The figures in parentheses are percentages of teachers giving the corresponding evaluation. The sum of the percentages in each row might not be exactly 100% due to rounding.

More important to this study is Part b of the question, which asks teachers how much the different sources contributed to their knowledge of how to use technology for teaching mathematics. Table 5.6 shows the distributions of teachers' evaluations of the contribution of different sources to the development of their knowledge of technology.

According to the table, we again can see that there are various sources from which teachers developed their knowledge of technology. However, in terms of the combined percentages of teachers' choosing positive evaluation: "very much" and "somewhat", the two most important sources for teachers to gain their knowledge of technology, as in the case of knowledge of textbooks, are teachers' "own teaching experience and reflection" (98.5%), and "informal exchanges with colleagues" (94.0%); "inservice training" (88.6%) and "organized professional activities" (79.1%) are also highly recognized; and "reading professional journals and books" (34.3%), "experience as student" (20.9%) and "preservice training" (20.9%) are reported to be the least important sources. Figure 5.4 presents a comparison of those sources.

Findings of the Chicago Study (I): Pedagogical Curricular Knowledge

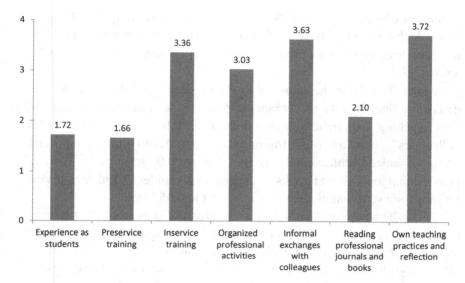

Figure 5.4 Comparison of the contribution of different sources to teachers' knowledge of technology

Note: By the ordinal scale in the figure, 4 = very much, 3 = somewhat, 2 = little, and 1 = no contribution.

From Figure 5.4, we can find that, overall, the order of these different sources for teachers to develop their knowledge of technology is, from the most important to the least important, "teachers' own teaching practices and reflection" (3.72), "informal exchanges with colleagues" (3.63), "inservice training" (3.36), "organized professional activities" (3.03), "reading professional journals and books" (2.10), "experience as student" (1.72), and "preservice training" (1.66), which, as expected, is consistent with the results revealed in Table 5.6.

Logistic regression models were again employed to analyze the collected data. The main result is included in Table 1F.2 of Appendix F.

According to the parameter estimates of the models shown in Table 1F.2, we can see that the order of importance of the sources to the development of teachers' knowledge of technology are "teachers' own teaching experiences and reflection" (G: -1.9138), "informal exchanges with colleagues" (E: -1.6493), "inservice training" (C: -0.9082), "organized professional activities" (D: 0), "reading professional journals and books" (F: 1.9272), "experience as student" (A: 2.7950), and "preservice training" (B: 2.9424). The order based on the above log-linear regression is the same as revealed by teachers' average ratings of the contribution of the sources (see Figure 5.4 above).

Meanwhile, from the values of Pr > Chi-Square in Table 1F.2, we can determine that, relative to "organized professional activities", "teachers' own teaching experiences and reflection", "informal exchanges with colleagues", and "inservice training" are significantly more important (than "organized professional activities") at the 0.05 level, and "reading professional journals and books", "experience as student", and "preservice training" are significantly less important at the 0.05 level.

As in the analysis of knowledge of textbooks, how teachers' teaching experience is associated with their evaluation of the contribution of different sources to their knowledge of technology was analyzed by comparing three different groups of teachers' average evaluations, and by applying the chi-square test to the detailed distributions of the numbers of teachers in the three groups having different evaluations of the contribution of each source to the development of their knowledge of technology. Figure 5.5 presents the comparison of the average evaluations.

According to the average evaluation shown in the figure, we find that there exist relatively large differences among those three groups for Source A ("experience as student") and Source B ("preservice training"), while the differences for the other sources seem small.

Now let us further examine the actual distributions of the numbers of teachers in the three groups having different evaluations, and employ the chi-square test to see if there exist statistically significant differences among those three groups of teachers' evaluations. Table 5.7 shows the result of the statistical processing of the data.

It is interesting to note that, largely consistent with the descriptive statistics presented in Figure 5.5, the results of the chi-square significant test reported in the Table 5.7 reveal that teachers across the three groups have statistically the same evaluations of the contribution of "inservice training", "organized professional activities", "informal exchanges with colleagues", "reading professional journals and books", and "own teaching experience and reflection" to their knowledge of technology. At the same time, they have significantly different views of the contribution of the other two sources: "experience as student" and "preservice training".

Findings of the Chicago Study (I): Pedagogical Curricular Knowledge

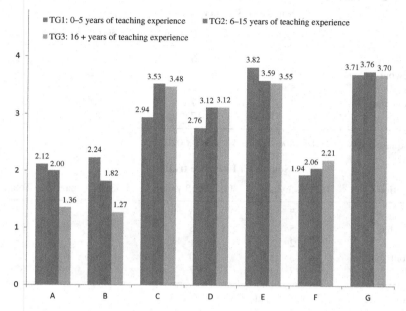

Figure 5.5 Comparison of the three groups of teachers' average evaluation of the contribution of different sources to teachers' knowledge of technology

Note: 1. For sources, A = Experience as student, B = Preservice training, C = Inservice training, D = Organized professional activities, E = Informal exchanges with colleagues, F = Reading professional journals and books, G = Own teaching experience and reflection.

2. The evaluations are shown by the ordinal scale in the figure, 4 = very much, 3 = somewhat, 2 = little, and 1 = no contribution.

Table 5.7 Distributions of the numbers of teachers among the three groups giving different evaluations of the contribution of each source to their knowledge of technology

	Source A: Experience as student			
Evaluation	TG1(0–5 years)	TG2(6–15 years)	TG3(16+ years)	Total
Very much	1	1	0	2
Somewhat	4	5	3	12
Little	8	4	6	18
No contribution	4	7	24	35
Total	17	17	33	67

Chi-square test*: $\chi^2 = 13.7867$ df $= 4$ $p = 0.0080^\dagger$

(*Continued*)

Table 5.7 (*Continued*)

Source B: Preservice training

Evaluation	TG1(0–5 years)	TG2(6–15 years)	TG3(16+ years)	Total
Very much	1	0	0	1
Somewhat	7	4	2	13
Little	5	6	5	16
No contribution	4	7	26	37
Total	17	17	33	67

Chi-square test*: $\chi^2 = 16.9406$ df = 4 p = 0.0020†

Source C: Inservice training

Evaluation	TG1(0–5 years)	TG2(6–15 years)	TG3(16+ years)	Total
Very much	7	10	18	35
Somewhat	5	6	13	24
Little	2	1	2	5
No contribution	3	0	0	3
Total	17	17	33	67

Chi-square test*: $\chi^2 = 10.2758$ df = 6 p = 0.1135

Source D: Organized professional activities

Evaluation	TG1(0–5 years)	TG2(6–15 years)	TG3(16+ years)	Total
Very much	3	4	13	20
Somewhat	9	11	13	33
Little	3	2	5	10
No contribution	2	0	2	4
Total	17	17	33	67

Chi-square test*: $\chi^2 = 5.7757$ df = 6 p = 0.4489

Source E: Informal exchanges with colleagues

Evaluation	TG1(0–5 years)	TG2(6–15 years)	TG3(16+ years)	Total
Very much	15	12	19	46
Somewhat	1	3	13	17
Little	1	2	1	4
No contribution	0	0	0	0
Total	17	17	33	67

Chi-square test**: $\chi^2 = 8.4845$ df = 4 p = 0.07754

(*Continued*)

Table 5.7 (*Continued*)

Source F: Reading professional journals and books				
Evaluation	TG1(0–5 years)	TG2(6–15 years)	TG3(16+ years)	Total
Very much	0	1	1	2
Somewhat	4	4	13	21
Little	8	7	11	26
No contribution	5	5	8	18
Total	17	17	33	67

Chi-square test*: $\chi^2 = 2.0715$ df $= 4$ p $= 0.7226$

Source G: Own teaching practices and reflection				
Evaluation	TG1(0–5 years)	TG2(6–15 years)	TG3(16+ years)	Total
Very much	13	13	23	49
Somewhat	3	4	10	17
Little	1	0	0	1
No contribution	0	0	0	0
Total	17	17	33	67

Chi-square test** $\chi^2 = 3.7850$ df $= 4$ p $= 0.4359$

* For each of Sources A, B, and F, Row 1 and Row 2 are combined for the chi-square test, for each of the original 4 × 3 tables produces three expected cell frequencies less than 1. The combined 3 × 3 tables nicely meet the common assumption of using the chi-square test.
** For each of Sources E and G, because all the cells in the last row are 0, the chi-square test cannot be directly applied. The results were obtained by applying the test to the rest of the table, excluding the last row.
† Significant at the 0.05 level.

Further applying chi-square tests to TG1 and TG2, TG1 and TG3, and TG2 and TG3 shows that for each of the two sources, there exist significant differences at the 0.05 level between TG1 and TG3 (for "experience as student", $\chi^2 = 11.8028$, df $= 3$, and p $= 0.0081$; for "preservice training", $\chi^2 = 14.4773$, df $= 3$, and p $= 0.0023$), and TG2 and TG3 (for "experience as student", $\chi^2 = 6.2417$, df $= 2$, and p $= 0.0441$;[8] for

---

[8]The results were obtained by combining the first two rows in the original 4 × 2 table to form a 3 × 1 table, and then using the chi-square test.

"preservice training", $\chi^2 = 7.3273$, df = 2, and p = 0.0256[9]), while there are no significant differences between TG1 and TG2 (for "experience as student", $\chi^2 = 2.2626$, df = 3, and p = 0.5197; for "preservice training", $\chi^2 = 2.5515$, df = 3, and p = 0.4661).

Also taking the numerical distributions of teachers' evaluation of those two sources into account, we can see that significantly more portions of teachers in TG1 and TG2 believed the sources contributed "very much" or "somewhat" to their knowledge of technology. More generally speaking, teachers with less years of teaching experience viewed those two sources more favorably than those with longer teaching experience.

In my view, the above result clearly reflects the fact that teachers with less teaching experience have been more exposed to using technology in their schools and colleges because they likely finished their school and college education more recently, and technology then was more available and advanced than before. This fact was also pointed out by some teachers during the interviews, which are reported below.

All of the nine teachers observed used or allowed students to use calculators in one or both of the classrooms observed, and one teacher also used computers in one of two classes observed. The interviews with these nine teachers reveal a pattern of how different sources contributed to those teachers' knowledge of technology, which is largely consistent with what was revealed from the questionnaire data discussed above.

Experience as student. None of those nine teachers described his/her experiences as a school student as a source for him/her to develop his/her knowledge of how to use technology to teach mathematics. If we notice that, according to Table 5.7, there are 14 (20.9%) out of the 67 teachers who answered in the questionnaire that "experience as student" contributed at least somewhat to their knowledge of technology, then it is likely that all the nine teachers interviewed belong to the remaining 79.1%.

Preservice training. Two teachers, each having less than five years' teaching experience, stated that their preservice training was helpful. Below are the dialogues between the interviewer (myself) and the teachers.

---

[9]The results were obtained by excluding the first row in the original table as both of the cells in the row equal 0.

> Interviewer: I noticed that you used a lot of calculators in the algebra class I observed yesterday. My question is how did you develop your knowledge of how to use technology? Here, it's calculator.
>
> TB1: That just came from — I was introduced to the TI-80s calculator when I was in graduate school. I learned how to use it there. I started with the 81, and then the 82 came out and I brought an 82. And then I just pretty much taught myself how to use it.

TC1 worked in the insurance industry for 15 years before switching to teaching three years ago. He emphasized the usefulness of his student teaching experience to develop his knowledge of how to use technology for teaching mathematics.

> Interviewer: I noticed that you asked them [students] to use calculators to check their answers and their calculations. How did you get to know how to use calculators?
>
> TC1: Oh, mainly business. When I first started, the calculators weren't as advanced as they are today. So I've been learning that when I student taught here. I taught two classes. I taught an Algebra class, which is a lot computer oriented. Those are classes they bring in here to work on computers. So that was really my real first experience with using technology in classroom.
>
> Interviewer: I see. So it is your student teaching experience.
>
> TC1: Yes. Student teaching experience is where I got most of my technology background.

It is interesting to note when the other teacher also with less than five years of teaching experience was asked if she developed knowledge of technology from her college or school experience, she replied: "No. Not at all. I never used any of the graphing calculators, none of the geometry software or anything in college."

Another six teachers interviewed also did not describe their preservice training experience as a source to their knowledge of technology.

Inservice training. Among the nine teachers interviewed, six reported inservice training as a source for them to develop their knowledge of technology. The following are excerpts from the records of the interviews with two teachers, one with nine years' teaching experience and the other with more than 20 years'.

> Interviewer: What kind of calculator do you use yourself?
>
> TA2: I usually use a TI-82, but I am comfortable with any of the TIs.

> Interviewer: OK. Then how did you learn how to use TIs?
>
> TA2: ... I've been to quite a few conferences or workshops on TI-86, TI-83, and TI-82. I've learned a lot of features there and how to use it. The conferences are a really good source of getting ways to use your technology in class. [For example.] After we saw a talk about using TI-83 for a parametric project, or polar project, I and my colleague, who's another precalc teachers, actually did that last year in our classes.

TC3's answer was also very clear in the aspect.

> Interviewer: I noticed in algebra class I observed, you asked students to check sin 30° using calculators. Did you use the TI-83?
>
> TC3. Right. TI-83, or TI-82. Whichever they want.
>
> Interviewer: When did you get to know how to use it?
>
> TC3: About six years ago, I went to a workshop at Lisle Junior High. It was a three day workshop. It was very intensive. They took the TI-81 calculator and taught us every button and every use in three days on the calculator. We went to our home school, and then throughout the year, we practiced it. ... and then every year, I go to the T^3 institute[10] and learn more [about] whatever the current calculator is.
>
> Interviewer: Did you attend seminars about how to use computers?
>
> TC3: Yes. That would be the ICTM[11] and T^3 two different sources.

As discussed earlier, inservice training did not play a major role in developing teachers' knowledge of textbooks. However, according to both questionnaire and interview data, it is not the case with teachers' knowledge of technology. To interpret that, I think the reason is that technology is relatively new and has been increasingly widely used in teaching activities; people, including mathematics educators, school administrators, and mathematics teachers, have recognized its importance, and there are a lot of training opportunities focusing on technology available to teachers.

Organized professional activities. As discussed in Chapter 3, by "organized professional activities" this study excludes those particularly designed for professional training. One way to distinguish "organized professional activities" from "inservice training" is that in the latter, there is a trainer(s), and teachers are the trainees, but in the former, teachers are more exchangers or learners (less formal than trainees).

---

[10] T^3 stands for "Teachers Teaching with Technology"; see on-line website: http://www.ti.com/calc/docs/t3.htm for more information.

[11] Illinois Council of Teachers of Mathematics.

Four of the nine teachers mentioned that they learned knowledge of technology from the professional activities they attended. For example, one teacher expressed that she did not attend any special training program outside, but learned how to use calculators in workshops organized by her department after school or during the summer. Another teacher said that he learned a lot from professional meetings he attended.

TC3 explained how a professional activity organized at school helped her to develop her knowledge of the use of computers with a particular piece of software, Sketchpad:

> What we did with the computers is some of my colleagues were working together with the Sketchpad. We had a study group. Our superintendent allowed us for our professional growth hours that we could get together. (Interviewer: When?) What we can do is any time on our own. We can make a proposal that this is what we want to study. Then we get together with our colleagues and study the geometry Sketchpad. We came in Saturdays and we practiced with Sketchpad and we talked about where do we think this will work in the classroom. What kind of lesson would be good? So it was the support of this school.

Informal exchanges with colleagues. During the interviews, six teachers pointed out that colleagues were their important source of their knowledge of technology. For example:

> TA2: [I learned how to use calculators] from my colleagues too. There's a lot of people here that know their calculators.
>
> TA3: I would say the biggest help was colleagues, because we have some people [in our department] that are very good with calculators.

In the geometry honors class observed, TA1 showed students how to use TI-82, TI-85, and TI-86 to convert 36.97° to the expression of degree, minute, and second. For TI-82, the key-use procedure was:

| 36.97 | 2nd | MATRX | 4:DMS | ENTER |

The output of the calculator was 36°58′12″. The interview was conducted after the class observation.

> Interviewer: ... How did you get knowledge of that?
>
> TA1: From colleagues.
>
> Interviewer: Did you attend any professional meetings [about how to use calculators]?

TA1: We have had some here at the school that I've gone to, so I've never gone to anything formal. But we had classes after school or during the summer I've gone to. And then, if a specific question comes up, we have a few people that are very strong in calculator, and I know that I can go to them. And they are always more than willing to sit down and help me with that.

Note that TA1, TA2, and TA3 are in the same school, and all of them consistently stated that there were some teachers in their department who were good with calculators. That reveals to some degree that the data collected are quite reliable.

In TC2's precalculus statistics class observed, the teacher showed students how to use calculators to graph step functions. Using TI-82, the following key-use procedure was instructed to graph the floor function $y = \lfloor x \rfloor$,

| MATH | NUM | 4. Int |

When the teacher was asked in the interview about how she obtained her knowledge of technology displayed in the classes observed, she answered:

Well, from two sources. I found out my own by experimenting with the machine myself, and the second I asked my colleagues. ... I found out either myself, or my colleagues. Like my colleagues, oh you can do the ceiling function in here no problem if you negate, and then I thought if it makes sense in terms of reflections...
I also attended professional seminars on the Internet. Interestingly enough, I learned how to use the Internet at the seminars, but I did not learn how to teach with the Internet. I learned that from my colleagues.

The teacher also said that she learned from her colleagues how to use the school computer lab to teach mathematics.

The last example here is from TC3, as she said:

This particular department is very, very knowledgeable, so you can learn a lot by just being here. We help each other out. If somebody gets a good idea, they share it. They don't hide it. So if somebody gets an idea how to use a graphic calculator or how to use Sketchpad, they pass the material onto everybody else in the department.

Reading professional journals and books. Only one teacher interviewed mentioned that reading professional journals might have been a source in her developing the knowledge of technology. All other teachers did not mention or consider reading professional journals as a source of their knowledge in this aspect.

Own teaching experience and reflection. Seven of the nine teachers described how, one way or another, their own teaching experience and reflection helped them to enhance their knowledge of technology. TB2 clearly explained how her own teaching experience and reflection (exploration) enabled herself to develop relevant knowledge about technology.

> Interviewer: Did you learn it [a computer software package] by yourself?
>
> TB2: Yes. It is me who asked the school to order it. After we got it, I sat down and learned it myself first and then wrote up a program for them (students) to follow.
>
> Interviewer: No special training?
>
> TB2: No.
>
> Interviewer: How did you get the knowledge of calculators you used in the classes I observed?
>
> TB2: I taught myself the graphing calculator while I was teaching. I also read through the manual, but it was mostly my own just playing with the calculator, my own experience.

Other teachers replied more briefly. For example, "My experience also helps", "I learned on the job", and so on.

Other sources. The other sources not discussed above include, first, a friend of a teacher, from whom the teacher also learned some knowledge of technology. Although the friend is not a colleague of the teacher in the same school, she is still a teacher. Therefore, in a broader sense, we can still treat her as a colleague of the teacher. Second, a teacher told that his previous (insurance) business experience also contributed to his knowledge of technology, which obviously belongs to the first component of teachers' knowledge of technology discussed earlier.

Table 5.8 summarizes the data from the interviews with the nine teachers.

In conclusion, from both the questionnaire and the interview data, we can see that there are various sources by which teachers develop their knowledge of technology. Overall, the most important sources are teachers' own teaching experience and reflection, their daily exchanges with their colleagues, and inservice training. Teachers' attending organized professional activities is a secondarily important source. Reading professional journals

Table 5.8 Sources of teachers' knowledge of technology (from interview data)

	A	B	C	D	E	F	G	Other sources
TA1				yes	yes			
TB1		yes			yes		yes	a friend
TC1		yes	yes					business experience
TA2			yes	yes	yes		yes	
TB2							yes	
TC2			yes		yes	yes	yes	
TA3			yes	yes	yes		yes	
TB3			yes				yes	
TC3			yes	yes	yes		yes	
Total	0	2	6	4	6	1	7	

Note: For sources, A = Experience as student, B = Preservice training, C = Inservice training, D = Organized professional activities, E = Informal exchanges with colleagues, F = Reading professional journals and books, G = Own teaching experience and reflection.

and books, experience as school students, and preservice training are the least important ones. Statistically, the contributions of those three kinds of sources to teachers' knowledge of technology are significantly different. Furthermore, experiences as students and preservice training played a significantly more important role in developing knowledge of technology for teachers with fewer years of teaching experience than for their senior colleagues, while each of the other sources has the same importance to teachers with different teaching experience in developing their knowledge of technology.

Knowledge of Other Teaching Resources

By teachers' knowledge of other teaching resources, this study mainly refers to their knowledge of how to use "concrete materials" for teaching mathematics. "Concrete materials", defined by Hiebert and Carpenter as "physical

Table 5.9 How often teachers used concrete materials to teach mathematics

	Always	Most of the time	Sometimes	Rarely	Never	Total
No. of teachers	0	11	48	10	0	69
% of teachers	0%	15.9%	69.6%	14.5%	0%	100%

Table 5.10 Distributions of the numbers of teachers giving different evaluations about the contribution of various sources to their knowledge of concrete materials

Sources	Degree of the contribution			
	Very much	Somewhat	Little	No contribution
Experience as student	1 (1.4%)	9 (13.0%)	23 (33.3%)	36 (52.2%)
Preservice training	3 (4.3%)	14 (20.3%)	23 (33.3%)	29 (42.0%)
Inservice training	12 (17.4%)	26 (37.7%)	18 (26.1%)	13 (18.8%)
Organized professional activities	10 (14.5%)	27 (39.1%)	19 (27.5%)	13 (18.8%)
Informal exchanges with colleagues	35 (50.7%)	25 (36.2%)	6 (8.7%)	3 (4.3%)
Reading professional journals and books	1 (1.4%)	18 (26.1%)	31 (44.9%)	19 (27.5%)
Own teaching practices and reflection	33 (47.8%)	28 (40.6%)	6 (8.7%)	2 (2.9%)

Note: n = 69. The figures in parentheses are percentages of teachers giving the corresponding evaluation. The sum of the percentages in each row might not be exactly 100% due to rounding.

three-dimensional objects" (1992, p. 70) and traditionally used by teachers to illustrate or to motivate mathematical ideas and operations (Begle, 1979), are explicitly treated in the NCTM *Standards* as instructional resources.

Question 17 of the questionnaire is about teachers' knowledge of concrete materials. Part a of the question investigates how often teachers used concrete materials in their classroom teaching. Table 5.9 displays the results from teachers' responses.

According to the table, it is clear that most teachers (69.6%) use concrete materials sometimes in their classroom teaching.

Back to the theme of this study, Part b of the question asks teachers how much the different sources contributed to their knowledge of how to utilize concrete materials for teaching mathematics. Table 5.10 shows how teachers evaluated those sources.

Following the same criteria we used above for teachers' knowledge of textbooks and technology (see Table 5.2 and Table 5.6) to analyze the data shown in Table 5.10, we can see that there are various sources for teachers to develop their knowledge of concrete materials for teaching mathematics. Besides that, in terms of the combined percentages of teachers' choosing positive evaluation: "very much" and "somewhat", overall teachers' "own teaching experience and reflection" (88.4%) and "informal exchanges with colleagues" (87.0%) are the two most important sources for them to develop their knowledge of concrete materials. "Inservice training" (55.1%) and "organized professional activities" (53.6%) are the secondarily important sources. "Reading professional journals and books" (27.5%), "preservice training" (24.6%), and "experience as student" (14.5%) are the least important ones.

Figure 5.6 shows a comparison of the contribution of different sources to teachers' knowledge of concrete materials, based on teachers' average evaluation in Part b of the question.

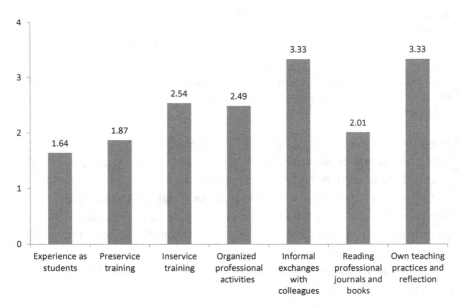

Figure 5.6 Comparison of the contribution of different sources to teachers' knowledge of concrete materials

Note: By the ordinal scale in the figure, 4 = very much, 3 = somewhat, 2 = little, and 1 = no contribution.

According to the average evaluation shown in the figure, we know that, overall, the order of the sources for teachers to develop their knowledge of how to use concrete materials for teaching mathematics is "teachers' own teaching experience and reflection" (3.33), "informal exchanges with colleagues" (3.33), "inservice training" (2.54), "organized professional activities" (2.49), "reading professional journals and books" (2.01), "preservice training" (1.87), and "experience as student" (1.64), from the most important to the least important, quite consistent with the result revealed in Table 5.10.

Once again, log-linear models were used to analyze the data collected from the questionnaire. Table 1F.3 in Appendix F displays the main result of the regression analysis via SAS.

Following the same method we employed for Tables 1F.1 and 1F.2 of Appendix F to examine Table 1F.3, we can see that, according to the parameter estimates in the models, the order of importance of the different sources to the development of teachers' knowledge of concrete materials are, teachers' "informal exchanges with colleagues" (E: -1.8788), "own teaching experience and reflection" (G: -1.8177), "inservice training" (C: -0.0902), "organized professional activities" (D: 0), "reading professional journals and books" (F: 0.9111), "preservice training" (B: 1.2824), and "experience as student" (A: 1.7641).

Again, we can note that the order of the importance of the different sources based on the above log-linear regression are the same as revealed by teachers' average ratings of the contribution of the sources (see Figure 5.6 above), except there is a difference between "informal exchanges with colleagues" and "(teachers') own teaching experience and reflection", which have the exactly same average rating, but are different between the parameter estimates in the logistic models, although the difference is quite small (0.06).

Moreover, from the values of Pr > Chi-Square in Table 1F.3, it is clear that, compared to "organized professional activities", teachers' "own teaching experience and reflection" and "informal exchanges with colleagues" are significantly more important at the 0.05 level, "inservice training" is at the same level without significant difference, and "reading professional journals and books", "preservice training", and "experience as student" are significantly less important at the 0.05 level.

110   Investigating the Pedagogy of Mathematics

With respect to the influence of teachers' teaching experience on the importance of the different sources to the development of their knowledge of concrete materials, I again use the questionnaire data to obtain the average evaluation of teachers in each of the three different groups of the contribution of each source to their knowledge of concrete materials, and employ the chi-square test to analyze the actual distributions of their evaluations. The results of the average evaluation obtained from the three groups of teachers are presented in Figure 5.7 below.

Figure 5.7 reveals that the biggest difference of the average evaluation among the three groups of teachers with different amounts of teaching experiences exists for Source B, "preservice training", between TG1 and TG3 with the value being 0.69, while all other differences range from 0 (between TG1 and TG3 for Source F, "reading professional journals and

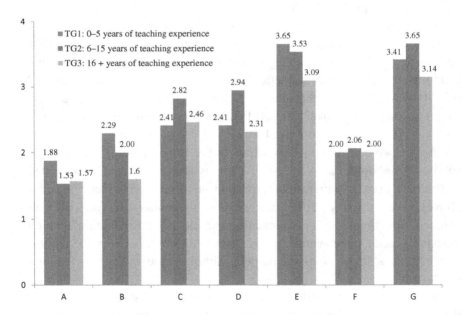

Figure 5.7 Comparison of the three groups of teachers' average evaluation of the contribution of different sources to teachers' knowledge of concrete materials

Note: 1. For sources, A = Experience as student, B = Preservice training, C = Inservice training, D = Organized professional activities, E = Informal exchanges with colleagues, F = Reading professional journals and books, G = Own teaching experience and reflection.

2. The evaluations are shown by the ordinal scale in the figure, 4 = very much, 3 = somewhat, 2 = little, and 1 = no contribution.

books") to 0.63 (between TG2 and TG3 for Source D, "organized professional activities").

The distributions of the three groups of teachers' evaluations, as well as the results of the chi-square tests, are displayed in Table 5.11.

From Table 5.11, we can see that, for the contribution of the following six sources to teachers' knowledge of concrete materials: "experience as student", "inservice training", "organized professional activities", "informal exchanges with colleagues", "reading professional journals", and "own teaching experience and reflection", there is no statistically significant difference of the evaluations among the three groups of teachers. However, there is a significant difference in teachers' evaluations of the contribution of the source "preservice training" across the three groups.

A further analysis using chi-square tests on TG1 and TG2, TG2 and TG3, and TG1 and TG3 leads to the result that the only significant difference exists between TG1 and TG3 ($\chi^2 = 15.1068$, df $= 3$, and p $= 0.0017$[12]), which is nicely consistent with what we found in Figure 5.7.

Taking the above statistical test results together with the numerical distributions in the table, we can generally conclude that junior teachers considered "preservice training" to have been more useful in developing their knowledge of concrete materials than their senior counterparts. It seems to me the main reason to explain the fact is that recent teacher education programs have paid more attention to teaching prospective teachers how to use "hands-on" and "manipulative" methods to teach mathematics, as they have been increasingly emphasized in mathematics education community in the last two decades.[13]

Unlike the case with technology, the nine teachers interviewed did not utilize concrete materials in their classrooms I observed, therefore I was not able to actually see how they might use concrete materials for mathematics teaching. The phenomenon is not quite unexpected.

---

[12]For TG1 and TG2, the results are $\chi^2 = 5.0357$, df $= 3$, and p $= 0.1692$; and for TG2 and TG3, $\chi^2 = 4.4863$, df $= 2$ (note: both cells in the first row are 0 and were excluded for the test), and p $= 0.1061$.

[13]For example, in more than 1,000 sessions presented at the NCTM 73rd annual meetings held in Boston, 1995, about 20% of sessions dealt with how to use "hands-on" and "manipulative" methods to teach mathematics. The two words were used most frequently in the titles of sessions. See more details in Fan (1995).

Table 5.11 Distributions of the numbers of teachers among the three groups giving different evaluations of the contribution of each source to their knowledge of concrete materials

	Source A: Experience as student			
Evaluation	TG1(0–5 years)	TG2(6–15 years)	TG3(16+ years)	Total
Very much	0	1	0	1
Somewhat	3	1	5	9
Little	9	4	10	23
No contribution	5	11	20	36
Total	17	17	35	69

Chi-square test*: $\chi^2 = 5.2597$ df $= 4$ p $= 0.2617$

	Source B: Preservice training			
Evaluation	TG1(0–5 years)	TG2(6–15 years)	TG3(16+ years)	Total
Very much	3	0	0	3
Somewhat	2	5	7	14
Little	9	7	7	23
No contribution	3	5	21	29
Total	17	17	35	69

Chi-square test*: $\chi^2 = 20.0333$ df $= 6$ p $= 0.0027$[†]

	Source C: Inservice training			
Evaluation	TG1(0–5 years)	TG2(6–15 years)	TG3(16+ years)	Total
Very much	1	4	7	12
Somewhat	8	8	10	26
Little	5	3	10	18
No contribution	3	2	8	13
Total	17	17	35	69

Chi-square test*: $\chi^2 = 4.7109$ df $= 6$ p $= 0.5814$

	Source D: Organized professional activities			
Evaluation	TG1(0–5 years)	TG2(6–15 years)	TG3(16+ years)	Total
Very much	3	4	3	10
Somewhat	5	10	12	27
Little	5	1	13	19
No contribution	4	2	7	13
Total	17	17	35	69

Chi-square test*: $\chi^2 = 8.9842$ df $= 6$ p $= 0.1745$

(Continued)

Table 5.11 (Continued)

Source E: Informal exchanges with colleagues

Evaluation	TG1(0–5 years)	TG2(6–15 years)	TG3(16+ years)	Total
Very much	12	9	14	35
Somewhat	4	8	13	25
Little	1	0	5	6
No contribution	0	0	4	3
Total	17	17	35	69

Chi-square test*: $\chi^2 = 9.181$ df $= 6$ p $= 0.1636$

Source F: Reading professional journals and books

Evaluation	TG1(0–5 years)	TG2(6–15 years)	TG3(16+ years)	Total
Very much	0	1	0	1
Somewhat	5	4	9	18
Little	7	7	17	31
No contribution	5	5	9	19
Total	17	17	35	69

Chi-square test*: $\chi^2 = 0.3812$ df $= 4$ p $= 0.9840$

Source G: Own teaching practices and reflection

Evaluation	TG1(0–5 years)	TG2(6–15 years)	TG3(16+ years)	Total
Very much	8	12	13	33
Somewhat	8	4	16	28
Little	1	1	4	6
No contribution	0	0	2	2
Total	17	17	35	69

Chi-square test*: $\chi^2 = 6.8438$ df $= 6$ p $= 0.3355$

* For each of Sources A and F, Row 1 and Row 2 are combined for the chi-square test, for each of the original 4 × 3 tables produce three expected cell frequencies less than 1. The combined 3 × 3 tables nicely meet the common assumption of using the chi-square test.
† Significant at the 0.05 level.

As people have found, while concrete materials are widely used in elementary school instruction, they are to a lesser extent used at secondary school level (e.g., see Begle, 1979, pp. 119–120). Besides, I believe that the increasing availability of calculators, computers, overheads projectors, and other teaching resources/tools further reduces, though does not totally eliminate, the necessity of using concrete materials for teaching mathematics. In fact, Table 5.9 shows that no one in the sample of this study always, and only 15.9% of teachers most of the time, used concrete materials in their teaching.

The fact that I did not see how those teachers would demonstrate the ways of using concrete materials in classrooms did raise difficulty for me to ask them relevant questions during the interviews, because I had no direct stimulants to use for asking questions, while teachers could not be given plenty of time to freely search their memories as all the interviews were time-limited. Nonetheless, all the teachers, except for one of them, were asked in one way or another the following question:

Do you use concrete materials in your teaching?

Six teachers answered "yes" or "sometimes". They were naturally asked further questions about how they obtained their knowledge of how to use concrete materials to teach mathematics; two other teachers were not asked the same further questions as one responded "rarely", and the other could not recall an example of her using concrete materials after responding "yes", instead she gave me some examples of using technology. For the teacher not asked the first question, I elicited twice at different times, "do you use some other teaching strategies I did not observe?", and he mentioned other strategies but not using concrete materials.

The most detailed response was obtained from TC1 (he even lent me a set of concrete materials, "Pentominoes").

> We will use concrete materials later. We haven't a lot this year. Last year in my class, when we were discussing probability and statistics, I taught them different dice games. So I brought in dice and used those as something to catch students' interest and then learn the mathematics behind it. (Interviewer: How did you know that kind of knowledge?) I learned how to play dice when I was a kid. . . . Last year I also attended one seminar at MMC[14] which was on that. . . . I used "Pentominoes" in my geometry class last year and I'll probably use them in the algebra class this year. (Interviewer: Is that one thing you learned from the seminar?) No, I learned that from a class I took in when I was a grad student . . .

Other teachers' answers are simpler and more straightforward. Table 5.12 summarizes their responses during the interviews.

---

[14] Metropolitan Mathematics Club of Chicago.

Table 5.12 Sources of teachers' knowledge of concrete materials (from the interview data)

	A	B	C	D	E	F	G	Other sources
TA1					yes		yes	
TB1	(note: questions not directly asked)							
TC1		yes		yes				experience as learner in informal environment
TA2	(note: concrete materials rarely used)							
TB2	(note: relevant answers not provided)							
TC2				yes				a friend; business experience
TA3					yes	yes		
TB3							yes	
TC3				yes				
Total	0	1	0	2	3	1	2	

Note: For sources, A = Experience as student, B = Preservice training, C = Inservice training, D = Organized professional activities, E = Informal exchanges with colleagues, F = Reading professional journals and books, G = Own teaching experience and reflection.

It is clear from the table that teachers have various sources to gain their knowledge of concrete materials for teaching mathematics. In addition to the seven possible sources listed in the questionnaire and confirmed from the questionnaire survey, the interviews revealed that teachers' experience as learners in informal educational environments, their friends, and their previous business experience could also inform their knowledge of concrete materials for teaching mathematics. It seems reasonable to expect that more sources might be identified if more teachers were interviewed.

However, it is unclear just from the table which sources are more important than others, because the sample of the interviewees about knowledge of concrete materials consisted virtually of only six teachers, and the result shown in Table 5.12 is less representative than that obtained from the questionnaire, as well as that shown in Tables 5.4 and 5.8 which were from all the nine teachers. That might explain why there is some inconsistency that can be found between the data obtained from the interviews and that from the questionnaire. For example, most obviously, in the interviews no one described "inservice training" as a source, but in the questionnaire it was chosen as the third most important source based on teachers' average evaluation of the different sources. Of course, there is also some consistency, for example, informal exchanges with colleagues were described by the most teachers to be a source in the interviews and were also believed to be one of the most important sources in the questionnaire survey.

I draw three main conclusions from this section. The first is from both the questionnaire and the interview data, and the other two primarily from the questionnaire data. That is, first, there are various sources by which teachers can develop their knowledge of concrete materials; second, comparatively, teachers' "informal exchanges with colleagues" and "own teaching experience and reflection" are the most important sources; "organized professional activities" and "inservice training" are the secondarily important sources; and "reading professional journals and books", "preservice training", and "experience as student" are the least important sources. Statistically, the contributions of those three kinds of sources to teachers' knowledge of concrete materials are significantly different. Finally, the length of teachers' teaching experience overall did not play a significant role in their viewing of the contribution of the different sources to their knowledge of concrete materials, except that "preservice training" received more favorable evaluations from junior teachers than from senior ones.

Summary of the Findings

This chapter reports the findings about how teachers developed their PCrK, which is further broken down into three components: knowledge of teaching materials, knowledge of technology, and knowledge of concrete materials. The following main findings are obtained through the analysis of the data collected from both the questionnaire and interviews.

1. There are various sources which teachers used to develop all the three components of their PCrK.

    Besides the above point, as concerns different sources investigated in this study, there are the following findings about the relative importance of different sources to each component of teachers' PCrK.

2. Teachers' own teaching experience and reflection, and informal exchanges with colleagues are the two most important sources for all the three components of teachers' PCrK.
3. Inservice training is one of the most important sources for teachers' knowledge of technology, and a secondarily important source for their knowledge of teaching materials and concrete materials; organized professional activities is a secondarily important source for all the three components.
4. Experience as student is a secondarily important source for teachers' knowledge of teaching materials, and one of the least important sources for their knowledge of technology and concrete materials.
5. Reading professional journals and books, and preservice training are the least important sources for all the three components of teachers' PCrK. Table 5.13 summarizes the above findings.

Table 5.13 A summary of the importance of different sources to the development of teachers' PCrK

Group of sources	Knowledge of teaching materials	Knowledge of technology	Knowledge of concrete materials
Most important sources	Own teaching experience and reflection Informal exchanges with colleagues	Own teaching experience and reflection Informal exchanges with colleagues Inservice training	Own teaching experience and reflection Informal exchanges with colleagues
Secondarily important sources	Inservice training Organized professional activities Experience as student	Organized professional activities	Inservice training Organized professional Activities
Least important sources	Reading professional journals and books Preservice training	Reading professional journals and books Experience as student Preservice training	Reading professional journals and books Preservice training Experience as student

Note: Compared to "organized professional activities", the sources in the first group are significantly more important at the 0.05 level, the sources in the second group have no significant differences in terms of importance, and the sources in the third group are significantly less important at the 0.05 level.

Finally, there are three main findings about the influence of teachers' teaching experience on the importance of different sources to the development of their PCrK.

6. The length of teachers' teaching experience does not affect their evaluation of the importance of each of the sources to the development of their knowledge of textbooks.
7. Compared with their senior colleagues, teachers with fewer years of teaching experience view their experience as students and their preservice training more importantly in the development of their knowledge of technology. But for all the other sources, there is no statistically significant difference.
8. Teachers with fewer years of teaching experience view their preservice training more importantly in the development of their knowledge of concrete materials than their senior colleagues. But no statistically significant differences exist for any of the remaining sources.

Chapter 6

Findings of the Chicago Study (II): Pedagogical Content Knowledge

According to Shulman, pedagogical content knowledge (PCnK) includes:

> for the most regularly taught topics in one's subject area, the most useful forms of representation of those ideas, the most powerful analogies, illustrations, examples, explanations, and demonstrations — in a word, the ways of representing and formulating the subject that make it comprehensible to others... [and] an understanding of what makes the learning of specific topics easy or difficult: the conceptions and preconceptions that students of different ages and backgrounds bring with them to the learning of those most frequently taught topics and lessons (Shulman, 1986a, p. 9).

He claimed that PCnK did not receive much attention in educational research before, but that it is very important for teachers to be able to do their teaching job.

As pointed out in the literature review of Chapter 2, numerous studies have been centered on PCnK since Shulman first introduced the term. Some researchers have substantially modified, expanded, or even, in my view, somewhat abused Shulman's conceptualization. For example, Cochran *et al.* employed a so-called constructivist view to modify their PCnK into "Pedagogical Content Knowing", which is "a teacher's integrated understanding of four components of pedagogy, subject matter content, student characteristics, and the environmental context of learning" (Cochran *et al.*, 1993, p. 266). Grossman defined PCnK to be "composed of four central components": knowledge and beliefs about the purposes for teaching a particular subject at a particular level; knowledge of students' understanding, conceptions, and misconceptions of particular topics in a subject matter; knowledge of curriculum materials for teaching particular subject matter and knowledge about both the horizontal and vertical curricula for a subject; and knowledge of instructional strategies and representations for teaching particular topics (Grossman, 1988, pp. 15–18). Under those loose and wide definitions, it is hard to see the original meaning of PCnK proposed by Shulman.

This study utilizes the NCTM *Standards'* conceptualization of PCnK, which basically follows Shulman's definition.[1] Briefly speaking, teachers' PCnK means their knowledge of "ways to represent mathematics concepts and procedures", which mainly includes two connected things:

1. Knowledge of what the "ways" are. That is, how many ways, either powerful or not so powerful,[2] are there to represent a particular mathematics idea? What are those ways?
2. Knowledge of how to choose the ways from the perspective of teaching. Namely, what are the advantages and disadvantages of those ways for classroom teaching? What difficulties might students have in trying to understand a certain mathematics idea? What are easier ways for them to understand?

Researchers have stressed that teachers need a good repertoire of PCnK. For instance, Shulman stated that "the teacher must have at hand a veritable armamentarium of alternative forms of representation" (Shulman, 1986a, p. 9). Colton and Sparks-Langer claimed that "an elementary teacher instructing mathematics should have a store of pedagogical content knowledge indicating that certain manipulatives are helpful for teaching concepts related to fractions" (Colton & Sparks-Langer, 1993, p. 47). More generally, the NCTM *Standards* proposed that "teachers need a rich, deep knowledge of the variety of ways mathematical concepts and procedures

---

[1] The *Standards* does not use the phrase "pedagogical content knowledge".

[2] In a strict sense, according to Shulman's description quoted above, PCnK is limited to "the most useful representation" or "the most powerful analogies, illustrations, examples, explanations, and demonstrations". However, what constitutes "the most useful or powerful" is undefined and, to me, is subjective. This study takes into account any kind of knowledge, regardless of whether it is wrong or right, powerful or not. See the definition of knowledge discussed in Chapter 3.

may be modeled, understanding both the mathematical and developmental advantages and disadvantages in making selections among the various models" (NCTM, 1991, p. 151).

Based on the above definition of PCnK, the present study investigated how teachers developed their PCnK; in other words, how did they get their PCnK for teaching mathematics? To address the question, the study collected original data from the questionnaire survey, classroom observation, and interviews, which followed the classroom observation, as described in Chapter 4. Reported below is an analysis of the data collected from these data sources and the findings of the study.

Analysis of the Questionnaire Data

Questions 20 and 21 of the questionnaire (see Appendix 1A) were targeted at teachers' PCnK. Because there are "unlimited" examples for PCnK within the discipline of mathematics (Brown & Borko, 1992, p. 212), it is impossible to ask teachers about their PCnK of all topics in a questionnaire survey. Moreover, it is also impractical to make up a few common topics that all the teachers have taught recently, as the sample of this study consisted of teachers teaching all kinds of mathematics subjects and using various textbooks. Therefore, both questions were designed to be open-ended, asking teachers first to recall two recent lessons that they taught that contained new mathematics topics, then identify, respectively, the new topics (Part a), the ways they represented the topics to students (Part b), and finally answer how they got to know the ways (Part c).

To stimulate teachers' memory and set an example for them to answer the questions, a sample was offered in the questionnaire before the two questions, which turned out largely helpful as the majority of the teachers responded to the two questions clearly. Overall, 58 (84%) teachers answered Question 20 and 54 (78%) answered Question 21.[3]

Topics and representations

A wide spectrum of mathematics topics were offered by the teachers to Part a of each of the two questions. According to a conventional classification of

---

[3] One teacher who filled in "see #20 [the above question]" was excluded here.

mathematics branches, 36% of all the topics given by teachers in the questions were algebraic, 21% related to probability and statistics, 25% to geometry, 8% to calculus, 7% to analytic geometry, and 4% to trigonometry. In total, there were 112 topics given by teachers, and 98 of them were different.

Some examples were: in algebra, "solving quadratics by factoring", "rationalizing denominators in radical expressions"; in geometry, "[the] Pythagorean theorem", "finding relationships or properties in parallelograms", "surface area of a rectangle solid", "volume of pyramids and cones"; in analytic geometry, "sketching certain polar equations", "to transform the equation of a circle to the standard form"; in probability and statistics, "counting principle", "independent and dependent events"; in trigonometry, "law of cosines", "relations of sides in special triangles"; and in calculus, "limits of sequences and series as $n \to \infty$", "evaluating $f'(x)$ using limits $\lim_{n \to \infty} \frac{f(x+h)-f(x)}{h}$".

Table 6.1 delineates the distribution of the numbers of the topics given by those teachers in different mathematics branches.

It should be noted that the classification used in Table 6.1 does not necessarily follow courses with the same name. For example, an algebra textbook might include geometric topics. Also, a topic might appear in different courses; for instance, the Pythagorean theorem can be taught in algebra, geometry, or trigonometry courses. For a very few topics which

Table 6.1 Distributions of the mathematics topics given by teachers in Questions 20 and 21

	Mathematics topics						
	Algebra[a]	Geometry	Analytic geometry	Probability & statistics[b]	Trigonometry	Calculus	Total
Q. 20	18	17	2	13	2	6	58
Q. 21	22	11	6	10	2	3	54
Total[c]	40 (34)	28 (24)	8	23 (19)	4	9	112 (98)

[a]Including Algebra I and II (or advanced algebra).
[b]Including combination and permutation.
[c]The figures in parentheses are the numbers of actually different topics, which are different from the grand totals as a few teachers gave the same topics in the corresponding columns. For example, three teachers gave the same topic, "factoring theorem" (Algebra), and two teachers gave the same topic, "volume of a sphere" (Geometry). All topics given in the columns without parentheses are different.

seem to fit more than one category, such as "the Pythagorean theorem", the author's discretion was used to make appropriate classification.

Examining teachers' responses to Part b of each question about "the ways to represent the topics" reveals that two teachers did not correctly understand the question really asked in Part b. They gave answers about their PIK (general teaching strategy)[4] but not PCnK (particular mathematical representation on a topic), though they must have used one mathematical way or another to represent a mathematics idea to students.

For Question 20, after giving the topic in Part a as "solving quadratics by factoring", one teacher (No. 10) answered in Part b: "Class discussion format. The students recognized that the quadratic could be factored so I just had to show them the last part of this new concept." The latter sentence actually implied that the teacher employed a certain way to introduce the concept to students, however, it did not unfold what the way was. The second teacher (No. 23) chose the topic as "limits at a point", and wrote in Part b: "[I used] a group learning process using a worksheet", which obviously points to a general teaching strategy, not particularly about teaching the topic. In other words, the answer did not reveal what PCnK about the topic he/she employed.

Similarly, for Question 21, the first teacher (No. 10), giving the topic in Part a as "solving rational equations", answered in Part b: "Lecture format, but students would brainstorm on how to get to the next step." The other teacher (No. 23), filling in the topic in Part a as "mathematical induction", explained the way in Part b: "Lecture followed by directed practice in groups." Again, both responses gave the format (strategy) of teaching but hid the substance (content) of teaching, which the question is really about.

The other 56 teachers for Question 20 and 52 teachers for Question 21 provided the targeted information. For instance, one teacher (No. 1), after giving one topic as "surface area of prisms", described the way he/she represented it to students:

[I] gave each group of students a box, [and] ask them to figure out how much paper is necessary to cover the box. They had to find the area of each side of the box and add them.

---

[4]Teachers' knowledge of a general teaching strategy is considered to be their PIK. See more discussions in Chapter 7.

Another teacher (No. 63) chose a topic as "sum of arithmetic series", and explained the following way, which is well known in mathematics:

$$1 + 2 + 3 + \cdots + 98 + 99 + 100$$
$$= (1 + 100) + (2 + 99) + (3 + 98) + \cdots$$
$$= \frac{100}{2}(100 + 1).[\text{then using}] \text{ generalization } S_n = \frac{n}{2}(a_1 + a_2)$$

Still another teacher (No. 38) gave a topic of "introduction of the concepts of tangent, sine, and cosine in right triangles", and the way of representation:

> We developed the concept using similar triangles, i.e., these ratios are the same for all angles of the same degrees. Use of SOH-CAH-TOA to remember [the concepts].

All the teachers' responses, except for the two teachers (No. 10 and No. 23) who misunderstood Part b of the questions and gave no PCnK as reported above, were included when I analyzed the data about the sources of teachers' PCnK in Part c, which is reported below.

Sources of PCnK

While Part a and Part b of each of Questions 20 and 21 were designed to provide stimulants and context, Part c directly attacks a theme of this study.

Most of the teachers who correctly gave the information in the first two parts also offered clear and code-ready explanations in Part c about how they got their knowledge of the ways to represent the topics to students. The following descriptions were given by the teachers quoted above to the topics and ways of representations aforementioned:

> Teacher No. 1: No one [told me the way]. I thought of it myself.
> Teacher No. 38: [I] taught this concept as a student teacher, which was my first exposure to SOH-CAH-TOA as a memory aid.
> Teacher No. 63: I was shown this derivation method in high school.

Other examples showing teachers had clear memories include: "[I learned the way] from a colleague who observed it at an NCTM workshop" (Teacher No. 41, topic: ANOVA); "[I got it from] NCTM national meeting in San Diego last year Paul Foerster's presentation"[5] (Teacher No. 20, topic: limits); among others.

The initial data analysis also suggests that, in addition to the seven sources discussed in Chapter 5, another major source for PCnK is "textbooks". For example, Teacher No. 30 described in Part b that he began with a "real-life problem" to introduce the topic of "interior and exterior of circles":

> A truck [is] driving through a tunnel. [The teacher asked] "Can the truck fit?" describing the tunnel as a semicircle with a given diameter, opening discussion about a truck with given height and width being able to fit.

In Part c, "how did you get to know the way?" he answered: "[It is from] personal teaching experience and reflection, along with concrete examples in the textbooks." Below I will include "textbooks" as a source.

Nonetheless, there were three teachers who answered "don't know" or "don't remember" to Part c of each question, and another teacher answered to Part c of Question 21 that "I learned this from a professor at Indiana University", but the statement is not clear whether it was through preservice training, inservice training, attending the professor's lecture in a meeting, or just personal contact, and so forth.

Table 6.2 summarizes teachers' responses to Part c of Questions 20 and 21, giving different sources by which they got their knowledge of the ways described in Part b.

From the table, we can notice that the distributions across the questions are very similar. (The result of the chi-square test for the eight sources is $\chi^2 = 2.6145$, $df = 7$, and $p = 0.9562$.) This in a sense also reveals that the data is quite dependable. Because there is no significant difference between the responses to the two questions, the analysis below uses the total responses of teachers.

Figure 6.1 illustrates the proportions of teachers who reported in Part c of the two questions that they got to know the ways of representation of the topics from the corresponding sources.

---

[5]The year is 1996, and the title of the presentation was "Calculus: The four concepts — Their definitions, meanings, methods, and application" (Session 103). See NCTM (1996).

Table 6.2 Sources of teachers' PCnK reported in the questionnaire

Sources[a]	A	B	C	D	E	F	G	H	Other[b]
Q.20 (n = 56)[c]	4	2	3	7	18	4	26	12	3
Q.21 (n = 52)[c]	2	1	3	5	18	2	28	9	4
Total	6	3	6	12	36	6	54	21	7

[a] A = Experience as student, B = Preservice training, C = Inservice training, D = Organized professional activities, E = Informal exchanges with colleagues, F = Reading professional journals and books, G = Own teaching experience and reflection, H = Textbooks.
[b] See explanations earlier in this section.
[c] A teacher might report that he/she learned his/her PCnK about a topic from several sources, therefore the number of teachers is the same as the number of topics, but the number of sources is greater.

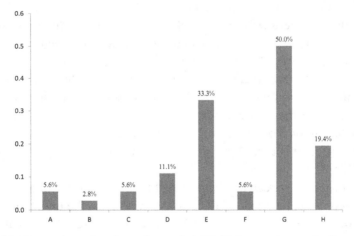

Figure 6.1 Proportions of teachers who got their PCnK about various topics from different sources

Note: 1. For the sources, A = Experience as student, B = Preservice training, C = Inservice training, D = Organized professional activities, E = Informal exchanges with colleagues, F = Reading professional journals and books, G = Own teaching experience and reflection, H = Textbooks.

2. The sum of the percentages is not 100% since some teachers reported that they learned their PCnK about certain topics from several sources. In other words, a percentage, say 50%, means that 50% of teachers learned their PCnK of the topics from Source G, but those 50% of teachers might also have learned from other sources.

As seen from Table 6.2 and Figure 6.1, there are various sources for teachers to get their PCnK of the topics they were teaching. Teachers overall developed most of their PCnK from their "own teaching experience and reflection" (50%) and their "informal exchanges with colleagues" (33.3%); they also obtained a considerable amount of their PCnK from "textbooks" (19.4%) and "organized professional activities" (11.1%); while the contribution of their "experiences as student(s)" (5.6%), "inservice training" (5.6%), "reading professional journals and books" (5.6%), and particularly "preservice training" (2.8%) are small.

Applying log-linear regression to the data, whose main result is attached in Appendix 1F (Table 1F.6), we can see that, according to the parameter estimates, the order of importance of the sources to the development of teachers' PCnK is the same as revealed by Figure 6.1, that is, teachers' "own teaching experiences and reflection" (G: 2.0794), "informal exchanges with colleagues" (E: 1.3863), "textbooks" (H: 0.6581), "organized professional activities" (D: 0), "experience as student" (A: -0.7538), "inservice training" (C: -0.7538), "reading professional journals and books" (F: -0.7538), and "preservice training" (B: -1.4759).

More importantly, from each value of Pr > Chi-Square in Table 1F.6, which is the significant level of the Wald chi-square test statistic (against the null hypotheses that the parameter equals 0), we detect that there are three groups of sources with different significance relative to "organized professional activities".[6] That is, compared to "organized professional activities", teachers' "own teaching experiences and reflection" and "informal exchanges with colleagues" are significantly more important at the 0.05 level; there is no significant difference between each source of "textbooks", "experience as student", "inservice training", and "reading professional journals and books" and the source of "organized professional activities" in terms of importance; and "preservice training" is significantly less important.

Now we consider how teachers' teaching experience is associated with the importance of different sources to the development of their PCnK. To be consistent with the analysis of teachers' PCrK discussed in Chapter 5,

---

[6]Note: we set $D = 0$ in the model.

Table 6.3 Numbers of teachers who answered the questions among the three groups

	TG1	TG2	TG3	Total
Q. 20	12	14	30	56
Q. 21	11	13	28	52
Total	23	27	58	108

Table 6.4 Teachers' responses across the three groups about the sources of their PCnK of the topics given in Questions 20 and 21

Sources[a]	A	B	C	D	E	F	G	H	Total
TG1 (n = 23)[b]	2	2	1	0	12	1	8	6	32
TG2 (n = 27)[b]	1	0	3	3	16	2	12	7	44
TG3 (n = 58)[b]	3	1	2	9	8	3	34	8	68
Total	6	3	6	12	36	6	54	21	144

[a] For the sources, A = Experience as student, B = Preservice training, C = Inservice training, D = Organized professional activities, E = Informal exchanges with colleagues, F = Reading professional journals and books, G = Own teaching experience and reflection, H = Textbooks.
[b] Here n is the total number of topics reported by teachers within the group in the two questions. See details in Table 6.3. Notice that a teacher might report that he/she learned his/her PCnK about a topic from several sources.

three groups of teachers were again used: TG1 consisting of teachers with 0–5 years' teaching experience, TG2 with 6–15 years', and TG3 with 16 or more years'. Table 6.3 shows the numbers of teachers who answered the questions within each of the three groups.

Table 6.4 tabulates teachers' responses about how they acquired their knowledge of the ways to represent the topics listed in Questions 20 and 21, broken down by different sources and teacher groups.

Applying the chi-square test to the data in Table 6.4 produces the result: $\chi^2 = 25.5914$, df = 14, and p = 0.02916, which rejects the null hypothesis that there is no association between the groups and the sources at the 0.05

Table 6.5 Teachers' responses across the three groups to the questions of their developing their PCnK from Source E

	Yes	No	Total
TG1	12	11	23
TG2	16	11	27
TG3	8	50	58
Total	36	72	108

level. In other words, there is a statistically significant difference among the three groups in their responses to how they developed their PCnK.

Further examining the numerical distributions of Table 6.4 reveals that the biggest difference across the three groups exists in Source E, "informal exchanges with colleagues". While in TG2 and TG1 teachers learned their PCnK from this source for 59.3% and 52.2% of the topics, respectively, the percentage is only 13.8% in TG3. Narrowing down the sources by removing Source E from the table, and applying the chi-square test to the rest of the data, we get $\chi^2 = 14.3168$, df $= 12$, p $= 0.2809$, which explains there is no statistically significant difference at the 0.05 level among those three groups for Sources A, B, C, D, F, G, and H.

Table 6.5 was derived from Table 6.4 in order to apply the chi-square test for Source E, which leads to $\chi^2 = 21.7915$, df $= 2$, and p $= 0.00002$. This verifies that there is significant difference among the three groups for Source E. Similarly, we can apply chi-square tests to detect the differences between TG1 and TG2, TG1 and TG3, and TG2 and TG3. The results reveal that, as we expect from the descriptive data in Table 6.5, significant differences exist between TG1 and TG3, and TG2 and TG3, with each p < 0.00001, but not between TG1 and TG2 with $\chi^2 = 1.3971$, df $= 1$, and p $= 0.2372$.

Regarding why there is a significant difference of the importance of Source E to teachers' PCnK among teachers with different lengths of teaching experience, I think the reason is that senior teachers would naturally give, instead of receive, more help to others especially to their younger colleagues, about how to represent mathematical topics to students, because they overall have more experience of teaching those topics. In other words, the two-way exchanges between senior and junior mathematics teachers are not equal; the juniors learned more from such activities.

In summary, from the questionnaire data of Questions 20 and 21, there are various sources from which teachers develop their PCnK. Overall, teachers learn more about how to represent mathematics topics to students from their own teaching experience and reflection and their daily exchanges with colleagues than from other sources. Textbooks, experience as students, attending organized professional activities, inservice training, and reading professional journals and books are the secondarily important sources. Preservice training is the least important source. Statistically, the contributions of those three groups of sources are significantly different. Furthermore, all of those sources have the same importance to teachers with different lengths of teaching experience, except for the source of "informal exchanges with colleagues", from which younger teachers learn significantly more than their senior colleagues.

Analysis of the Interview Data

Unlike the questionnaire survey, which specifically focuses on the core of teachers' PCnK, namely their knowledge of the ways to represent certain mathematics topics, the classroom observations with nine teachers provided wider opportunities for this researcher to identify the various kinds of PCnK those teachers have. In addition to forms of representation of mathematics topics, what Shulman calls "[the most powerful] analogies, illustrations, examples, explanations, and demonstrations" (Shulman, 1986a, p. 9) in his conception of PCnK also received attention during the observations and was identified after the observations.

The interviews with the nine teachers were based on the classroom observations. As in the cases of the other components of teachers' pedagogical knowledge, there were always more pieces of knowledge that could be identified from the classroom observation than could be asked about their sources during the interview, because on the one hand teachers' knowledge permeated their teaching activities throughout a class period, and on the other hand, all interviews were time-limited and included questions about many aspects of teachers' pedagogical knowledge, not just PCnK. Therefore, I decided for the interview to focus on each teacher's three to five identified pieces of PCnK that were used to particularly introduce the new topics[7] of the lessons to students in the classes and ask them about how they developed such kinds of knowledge, which I believe were most important based on the definition above and also easy to communicate with the teachers. For each teacher, the focus was usually on only one class, so teachers did not often need to shift their attention and memory between two classes, and questions could more naturally follow one another. Nonetheless, some attention was also paid to the other observed class.

In most cases, teachers' answers were clear and specific, and provided detailed and contextualized information to examine the issue. In a very few cases, teachers did not provide clear information or went off the topic, which were not further coded and hence excluded from the results reported below.

Interview with TA1

TA1's algebra class observed was devoted to the introduction of "the real number line", Lesson 2.1 in the textbook, *Algebra 1: An Integrated Approach* (Larson, Kanold, & Stiff, 1995a, pp. 62–67). According to the textbook, the lesson has two goals: (1) graph and compare real numbers using the real number line, and (2) find the opposite and the absolute value of a real number.

While basically following the text to represent the topic to students, TA1 emphasized more the first goal, the number line, and its use to compare negative numbers. She omitted the last example in the text which serves the second goal. When asked what the most difficult or important knowledge for students to learn was in this lesson, she gave the following response which reveals that she redesigned the way to represent the topic based on her past teaching experience and reflection.

> For the algebra class, opposite and absolute value will keep coming up. We'll keep talking about it. And so in the past, I found myself — and this is an experience thing — when I taught the section the past couple of years, the number line part, I just kind of — let's graph a couple of number lines. Let's compare them [numbers]. We're done. And then I really measured on opposite and absolute

---

[7] As expected, most teachers observed also spent a considerable amount of time on reviewing topics previously taught.

value. But we continued to do opposite and absolute value throughout the year. And then I found the students, when they're talking about negative numbers, getting confused on which one's less and which one's more. So I thought this year I was going to try it, experience-wise. I know it's happened in the past. This year, let's try it so that I spend even more time on doing a comparing thing, so they get a conceptual understanding of how the number line works, how negative numbers compare to each other, how fractions compare to one another. And then, as we go throughout the course, we'll keep talking about opposite and absolute values. We'll keep reviewing over those.

When asked a more detailed question, "Why did you emphasize how to write the numbers from the number line in decreasing order or in increasing order?", she responded:

Yes. Because of my past experience. Remember I told you that in past years, they had problems [in that]. In fact, it was very interesting, because the next period I taught that same lesson. And you know the back page where it had some positives and then it went to the negative numbers? (Interviewer: Yes.) They took the longest time figuring out what the biggest negative number was to get them started. It was really interesting. It made me feel like, OK, maybe this is a worthwhile activity.

TA1's geometry class observed was for the introduction of "Measurement of Segments and Angles", Lesson 1.2 in *Geometry: For Enjoyment and Challenge* (Rhoad, Milauskas, & Whipple, 1991, pp. 9–17). An analogy she used for the relationship among degree, minute, and second of an angle was hour, minute, and second of time, that is, using 1 hour = 60 minutes and 1 minute = 60 seconds in time measurement leads students to learn that 1 degree = 60 minutes and 1 minute = 60 seconds in angle measurement. When asked, "Did you learn that from the textbook?" TA1 answered:

I don't even know, to be honest with you, if the textbook says that.... I just know — and I think I also used different measurement, where we have meters, we have centimeters, [and] we have millimeters, you know. It gets them to understanding, because all they've pretty much been exposed to is degrees. So their first knowledge is, well, why do we have to have anything else? Degrees works. Why do we have to have — so, if I use comparison to them, something that they're used to, then it helps them understand a little bit more. (Interviewer: How did you know that?) I think again, it's just I look for comparisons of things when I'm thinking about things. I can get above a due amount of experience. And if I can have somebody say you do this like you do this, or you have this just like the reason you have this, it helps you get the bigger picture. I don't know if I got that from anywhere else.

According to the teacher's memory, it seems she got the idea mainly through her teaching experience and reflection.

Interview with TA2

The goal TA2 clearly put on the chalkboard in the algebra class observed was "to read graphs and analyze data", slightly different from the textbook used, which includes both "graphs" and "tables"[8] (see Lesson 1.8 in *Algebra 1: An Integrated Approach*, authored by Larson, Kanold, & Stiff, 1995a, pp. 49–54).

TA2 mainly used a series of questions, attached in Appendix 1G, to introduce the topic to students. He first let students do Questions 1–6 by themselves; after giving a brief explanation of those six questions, he demonstrated how to solve Questions 7–10 using the format of lecture; finally, he assigned students to do Questions 11–14 as in-class exercises. The following conversation reveals in more detail what PCnK TA2 used and how he developed such knowledge.

> Interviewer: OK. For the algebra class, you asked your students to do Questions 1 to 6. It's about bar graphs; and then you yourself explained Questions 7 to 10. Why did you make such a difference?
>
> TA2: OK. Because having taught the class a couple of years ago, I know that in middle school — most of them are freshmen, they have done bar graphs before. And they've done them pretty well. And they don't need a lot of instruction on that. But they haven't done frequency distributions and line plot. That's something that's sort of new in their book. So, in that case — it's not hard, but it's different. And so, I do one together. I sort of have an idea of what they already know how to do and what they haven't done before.
>
> Interviewer: Well, so because you know that. But how did you get to know that? Your experience?
>
> TA2: Yes, because having taught it the year before, or two years ago, the kids will tell you, we've done this already. They will come right out and say, we did this last year.
>
> Interviewer: So in the first year you taught it, you used a different way?
>
> TA2: I probably did a lot more directive teaching and less them doing it on their own when I started. I didn't know what they knew, so I assumed they knew very

---

[8] According to the teacher, "tables" were omitted because they were covered in the previous year.

little. Now I have a better understanding of what they actually already do know and should be able to do.

During his demonstrating how to answer Question 7, TA2 reminded students not to miss a number or put down the same number twice, a mistake they often made.

> Interviewer: OK. Now is a question about contents. In the algebra class I observed, you emphasized to students that they need to check the numbers, so they don't miss numbers or put a number twice. Why did you do that?
>
> TA2: Well, one of the things in terms of that is, I'm trying to train them to be in the habit of looking for their own mistakes in all kinds of problems, solving equations, the whole thing. So I tried to almost incorporate it into every lesson so that they're constantly being reminded that they have to check themselves. In terms of that specific problem, I just know from my own experience of having made frequency distributions, that if you go too fast, it's very easy to miss a number or write it down twice.
>
> Interviewer: From your teaching experience?
>
> TA2: Right.

In the precalculus honors class observed, which introduced, "Synthetic Division: The Remainder and Factor Theorems", Lesson 2-2 in *Advanced Mathematics: Precalculus with Discrete Mathematics and Data Analysis* (Brown, 1997), TA2 emphasized the importance of The Remainder Theorem. He explained he learnt this from the textbook as well as his previous teaching experience.

> Interviewer: In the precalculus class I observed, you told students that The Remainder Theorem is very significant, very important. How do you know that? Did someone tell you or textbooks tell you?
>
> TA2: I think mostly that comes from the textbook and also from having done the problems last year and the year before, that if a student doesn't understand the remainder theorem, they're going to have trouble dividing, looking for factors, factoring, finding roots of equations. That it's a really important element to a lot of topics that follow. So, from the textbook and from experience. Both.

A notable piece of TA2's PCnK demonstrated in the precalculus class is that, after introducing how to divide a polynomial by a binomial x − a using standard synthetic division, he presented the following example to explain how to use a similar procedure to divide a polynomial by a quadratic, a special algorithm rarely seen in mathematics books.

Example: Divide $x^5 - 2x^4 + 3x^3 - x^2 + 2x - 5$ by $x^2 - 2x + 3$

```
 2           -3  |   1    -2    3    -1    2    -5
      ×(-3) →               -3    0     0    3
      ×2 →                2     0     0    -2
                    ─────────────────────────────
                    1     0     0    -1    0    -2
```

The quotient is: $\quad 1x^3 + 0x^2 + 0x + -1$

The remainder is: $\qquad\qquad\qquad\qquad\qquad 0x + -2$

Below is the conversation with respect to this method:

Interviewer: And in the second class I observed, you introduced another kind of knowledge of synthetic division to divide a polynomial by a quadratic. It's not in the textbook.

TA2: Right. Correct.

Interviewer: Why did you decide to add that?

TA2: Well initially, a colleague of mine, [the name][9] showed me how to do it. And it started off, it was a topic that we used to teach our math team students. But he and I were sort of discussing it and we both sort of thought we could expect our students to do this. It's not that hard. Once you sort of learn the algorithm. And it's worth teaching. It allows you to go farther in terms of being able to divide polynomials. And it's just something we've always sort of added in. Now we won't — in terms of a test, we will put a question like that that's not too difficult, and mostly look for whether they can do it. Nor can they do all the little tricks and trades and difficult part of it, but the basic concept. But that's a basic concept that can be very useful to students.

Interviewer: I see. So you developed that idea from your colleague?

TA2: Yes. Both of us. We kind of did that together.

The conversation suggests that TA2 learned this special PCnK from one of his colleagues.

Interview with TA3

TA3's response about her PCnK focused on the first observed class, a calculus class, in which the new topic she represented to students was the four

---

[9]The actual name is not disclosed here for purposes of anonymity.

"basic differentiation rules", which is part of Lesson 2-2 in *Calculus with Analytic Geometry* (Larson, Hostetler, & Edwards, 1994, pp. 115–126). The other part of the lesson is about "rates of change", which the teacher, not following the order of the contents in the textbook, had already introduced before the class observed. She explained how she developed such an idea in the interview.

> Interviewer: I found "rates of change" is introduced later in the textbook. You introduced earlier. Why did you do that?
>
> TA3: Yes, because when you introduce derivative, which was introduced early, a derivative is a ratio of rates of change. And I don't want my students to just do things mechanically. I want them to understand what it really means. So I try to do the mechanics with the geometry of it. I try to do mechanics and geometry together. If you're going to talk about derivatives in terms of geometry, you have to talk about them in slopes of lines, so you have to talk about rates of change. So logically they go together. I guess you could say I am always trying to introduce ideas before we really get to them in the textbook.
>
> Interviewer: Why did you know it?
>
> TA3: I guess again experience and reflection. I mean, I've taught this before. And when you get to something, I say to myself, why I didn't say this before? It would have made this so much easier. So the next time I do it, I do it that way.
>
> Interviewer: So the first time you did it maybe–
>
> TA3: I may not like the way I do it the first time. Sometimes I do like the way I did it the first time. I mean, I always think about it. I don't just do things the way the book does it. I don't ever just do things just because the book does it. I look at the book. I decided if I like the book. If I really like the book, I use it. If I don't like the book, I try to find a better way to do it.

Although the textbook contained detailed proofs for the four rules shown here, TA3 only presented students the proof for the second rule:

$$\frac{d}{dx}(c) = 0;$$

$$\frac{d}{dx}(x^n) = nx^{n-1};$$

$$\frac{d}{dx}(cf(x)) = cf'(x);$$

$$\frac{d}{dx}[f(x) \pm g(x)] = f'(x) \pm g'(x).$$

The interview below illustrates what kind of PCnK she had and how she developed such knowledge.

> Interviewer: In the calculus class, you introduced four short-cut rules, but you only chose one to prove. I mean, that's the second one.
>
> TA3: Yes.
>
> Interviewer: You didn't give them the proof for the rest three. Why did you choose the second one?
>
> TA3: Because the first one was the derivative of a constant. It's obvious that if it's constant, it has no change. The three and fourth properties — derivatives are limits. Those are properties of limits. So if you understand the properties of limits, those properties are directly from the properties of limits. Therefore, we don't need to prove them.
>
> Interviewer: Yes. So are you sure the students will understand them easily?
>
> TA3: Yes.
>
> Interviewer: How can you be sure? I mean, how do you know? That's what I am interested in.
>
> TA3: I choose the ones that they have the most trouble accepting. If the ones that aren't obvious — they'll look at certain things and they'll go, well, that makes sense to me. And so, those I don't prove. But the ones that don't make sense or they'll go, well, how did you get that? I don't really understand that. Those are the ones I try to prove, so that they realize that there is a valid mathematical proof behind it.
>
> Interviewer: Can I say your past teaching experience helped you to make such a decision?
>
> TA3: Yes. But also my common sense. If it's obvious to me, it's probably obvious to them. And if it's not obvious to me, it's certainly not obvious to them.
>
> Interviewer: But the fourth rule here seems a little bit hard for me.
>
> TA3: Well, it was the property of limits. So if it's a property of limits — and we did it before with limits. See, derivatives are just limits. And so, if derivatives are limits, then all the properties that apply to limits should apply to derivatives. But the second one wasn't a property of limits. It was new. So I had to prove it.

TA3 also used different examples from the textbooks to represent the rules. When asked why she believed those examples would be helpful for students to understand the topics, TA3 pointed out it was because of her own teaching experience and reflection.

TA3's answer to the last question here generally deals with her PCnK of both the calculus class and geometry class, namely, in terms of Shulman, "an understanding of what makes the learning of specific topics easy or difficult" (Shulman, 1986a, p. 9).

> Interviewer: In the classes I observed, what's the most difficult in knowledge or concepts?
>
> TA3: The most difficult thing is students are very good at grinding out numbers or answers. It is very difficult for them to tell you exactly what those answers mean. What do they have? And that is what's difficult for them. You get an answer. What is it? You get a number. What is it? What does it mean?
>
> Interviewer: That's for all classes?
>
> TA3: Yes. But sometimes I think, when you're doing geometry, and you're finding an area of a triangle, and they get a number, it's easy for them to say, well, that number is the area. But in calculus, some of the things are a little more obscure. So it's harder for them sometimes to keep in mind what it is they really found. And so, I keep asking them. I keep asking them, tell me in words what you found. Draw me a picture of what you found. What do you have?
>
> Interviewer: So that's also from your experience?
>
> TA3: That's from experience.

Interview with TB1

The interview with TB1 about his PCnK mainly focused on the first class observed,[10] a geometry class, in which an important concept, "perpendicularity", was introduced. The textbook used is *Geometry: For Enjoyment and Challenge* (Rhoad, Milauskas, & Whipple, 1991, Lesson 2-1, pp. 61–65). Three kinds of PCnK were identified after the classroom observation, and questions about their sources were raised during the interview.

Different from the approach used in the textbook, which first directly gives the definition of perpendicularity, TB1 first asked students to fold a piece of paper on top of lines, draw a segment, and measure the angles. When asked how he designed the hands-on approach to the concept, he answered it was from his first-year teaching experience and reflection, not by other colleagues or some other sources.

---

[10] Two questions about the second observed class, an algebra one, were also asked, but the teacher's answers were not clear enough to be coded. For example, when asked "How did you know it [a point made in the algebra class] is important to let students know?", the teacher answered, "You see it a lot". All unclear responses are not included in this section.

After giving the definition of perpendicularity, TB1 stressed to students that one should not assume that an angle is a right angle just because it appears to be so in a picture.

> Interviewer: You asked that they don't assume perpendicularity from a picture. I mean, why did you say that's important?
>
> TB1: Well, because sometimes things aren't as they appear. And you should always have facts behind you before you assume something, either with given information or actual measuring.
>
> Interviewer: You mean this is one mistake students are easy to make?
>
> TB1: They'll see a picture that looks like a right angle and immediately... Interviewer: How did you know that? Why do you know students will make such kind of ... TB1: Past teaching experiences. They always like to assume things.
>
> Interviewer: So through your own experience you know that?
>
> TB1: Yes. It's probably also going back to when I was a geometry student. I'd probably always assume that too. I saw an angle. It looked like it was a right angle.

One subtle distinction of language TB1 made for the concept was that he told students that they should say perpendicular lines form a right angle, but not perpendicular lines equal ninety degrees. The conversation below reveals TB1 had a clear PCnK base for that point. Namely, he understood that it is easy for students to make mistakes.

> Interviewer: I also noticed when you introduced the definition of perpendicularity, you emphasized that when two segments or lines or whatever are perpendicular, they form right angles. Why did you emphasize that?
>
> TB1: Because they have a tough time understanding what happens when we're talking about these things. And even today if you would have come back, someone said perpendicular segments equal ninety degrees. So I think it's important for me to say they don't equal ninety degrees, but they form angles which are right angles.
>
> Interviewer: Again, I mean how did you know that?
>
> TB1: I think it goes back to when I was learning geometry what I did to make things easy for me. Most of the things that I do in class in terms of helping them out are things that I did as a student that helped me out. And I think the more times or the different ways they can see a concept, they're going to pick one particular way that would help them out.

It is clear that TB1 developed the knowledge from his experience as a school student.

Interview with TB2

The interview with TB2 focused on the observed algebra class, in which the teacher introduced students to "Quick Graphs of Linear Equations", Lesson 2-3 in *Algebra 2: An Integrated Approach* (Larson, Kanold, & Stiff, 1995b, pp. 78–85). Three methods were presented to graph a linear equation: using a table of values; using intercepts; and using the slope-intercept form.

Mathematically, graphing a linear equation only needs to have the values of two points from the equation. However, TB2 asked students to find values of three points to graph a linear equation.

> Interviewer: I noticed when you asked a student to graph the linear equation [$2x + 8y = 8$], you said "mathematically two points are needed, but it is better to have three points." Why [have] you emphasized 3 points?
>
> TB2: The day before that presentation, or actually it was probably two days before when we started the chapter. We started graphing lines and we did it using a table and I asked the class "how many points do we need to determine a line?" And they said two; they said two and I said, "okay, how many points should we suggest that we graph on this line?" and they of course then said "two". I said "but, how do you know if you've made a mistake? Will it be easier to tell if you've made a mistake if we plot three points?" And that's when they said "oh, okay". Then they understood what I was talking about. Then I said "in agreement then we are gonna agree when we plot out a line that we will have three points."
>
> Interviewer: Yes, that's like a check, double check?
>
> TB2: Yes, exactly, it was a check thing.
>
> Interviewer: When I was a student, my teacher did not tell me that. Why did you tell them? How did you get to know that?
>
> TB2: It was to hopefully to prevent some errors. It was based on seeing errors in the past, in my teaching experience of having them plot two points and then just draw the line through without thinking about it. So, hopefully to avoid some errors.
>
> Interviewer: So, this is from your teaching experience?
>
> TB2: Yes.
>
> Interviewer: I mean you did not get it from other teachers? A book or other teachers?
>
> TB2: No.

When students had difficulty using the intercepts to graph $y = 6x+18$ as the y-intercept, 18, is too high to be located in the x–y plane, TB2 introduced "rise over run" method, that is, after finding the x-intercept (−3, 0), move six units up and one unit right to locate a second point (−2, 6) to graph the equation. When asked how she got to know this alternative method, TB2 answered it was partly from her experience as a student and partly from the textbook she used for the class.

The following conversation is about TB2's knowledge of the most difficult things for students to learn in the algebra course.

> Interviewer: What do you think is the most difficult knowledge or concepts to learn in the algebra class?
>
> TB2: In solving what we call literal equations, where you have to solve for a variable given other variables.
>
> Interviewer: How did you know?
>
> TB2: Because, we just spent — this is Chapter 2 — in Chapter 1 we spent a couple of days talking about solving these kind of equations. So, I knew to anticipate that there would be some difficulty, because they did have a lot of difficulty with that.
>
> Interviewer: It is still from your teaching experience?
>
> TB2: Yes.

The teacher also said that she would think that students had difficulty in the calculus class, which I also observed, associating the definition of the derivative with actual problems, and some had a difficult time with the simplification of the algebra, the rational functions. However, she did not definitely explain how she formed such thoughts.

Interview with TB3

Both of TB3's classes observed were geometry, but the first class was for lower students and the second for accelerated students. The textbook for both classes was *Geometry: For Enjoyment and Challenge* (Rhoad, Milauskas, & Whipple, 1991). As the second class basically followed the textbook and used its questions to introduce proof, the interview about TB3's PCnK focused on the first class.

The first class was introducing the concept of "perpendicularity" (Rhoad, Milauskas, & Whipple, 1991, Lesson 2-1, pp. 61–65). Like TB1, TB3 also reminded students that they should not assume an angle is a right angle just from its appearance. When asked, "Why do you know that's important? Or why do you know students will easily make such a mistake?" TB3 responded,

> From my experience. They always want to assume that it's been drawn perfectly to scale. The textbook says that also. In the textbook, in this particular book, they have a list of what to assume and not to assume. So I was just reminding them that we had that back on page 19, what you should assume when you look at a diagram and what you should not assume.

TB3 gave the following three descriptions, written down on the chalkboard, to help students to grasp the concept of perpendicularity:

D. *Line, rays, or segments that intersect at right angles are perpendicular.*
A. *If lines, rays, or segments intersect at right angles, they are perpendicular.*
B. *If lines, rays, or segments are perpendicular, they intersect at right angles.*

The textbook only gives definition D. A discussion occurred in the interview with the intention to know how TB3 developed such an idea to represent the concept that way. At first TB3 could not answer the question.

> Interviewer: I observed that when you introduced the perpendicularity you wrote down three sentences. I mean, you could use the one in the book. So why did you give [them] three?
>
> TB3: Why did I write the others? Because the definition in the book was just a simple sentence: subject, verb, object. We study conditional sentences. If this, this, this, then this is your conclusion. Every sentence can be put into if-then form. If it's not a definition, the converse is not true. Not necessarily true. But if it's a definition, the converse is true.
>
> Interviewer: So that's from your own experience or you learned that from where or from whom?
>
> TB3: I don't know where I learned it.
>
> Interviewer: Did someone tell you that?
>
> TB3: No.
>
> Interviewer: Maybe from your own experience?
>
> TB3: I probably learned it from my first textbook I taught, because in the textbook in geometry, we make a big deal when they start proofs. And they slowly, slowly start proofs. They make a big deal of showing you conditions in geometry. Not necessarily higher up, but in the basics of geometry. And that's why, from my

Findings of the Chicago Study (II): Pedagogical Content Knowledge 143

experience in the textbook, it is so easy for me to say, this must be in the "then" part of the definition you choose. Because you know what? A lot of students say, I don't know what to pick for the reason. And that's how I demonstrate it to them. So that's why every time I give them a definition, I show them that the converse is true. But when I give them a theorem, it's not necessarily true.

According to the coding system of this study, it seems most reasonable that TB3 formed such an idea through her long-time teaching experience and reflection.

TB3 used three ways to represent the concept of perpendicularity to students: oral explanation, displaying in written form on the screen, and demonstrating by a hands-on activity: asking a student to hold a ruler to make a right angle. The interview proved that they are not casual teaching behaviors. In fact, TB3 explained that she intentionally designed the ways herself and traced back the knowledge base to professional training she received on psychology.

> TB3: I try to give as many different ways of visual as possible. Some people like to see it, like on TV, on the screen. Other people like to have an object in their hand.
>
> Interviewer: I see. So those ways were designed by yourself?
>
> TB3: As many ways as I can demonstrate, if I can put it on the blackboard, on the screen, and if I have a proper visualization. I try to do as many different presentations of the same thing.
>
> Interviewer: How did you know to design such a way, I mean, asking the students to make up angles at ninety degrees? Did you figure that out by yourself?
>
> TB3: I don't know why other than when I see a word like "perpendicular", I know that I can write it down. Not everybody can read and immediately know. So then I try to do a visual. And that's where the rulers and the — so I try to do what they call visual, as well as orally. I try to do everything.
>
> Interviewer: Yes. When did you become aware visual is helpful?
>
> TB3: Probably at early, early on when I was — you know how I told you every summer I go to school? Way back when, they had this on Piaget. And they had right brain, left brain. I happened to go to a couple of those classes.
>
> Interviewer: I see. Seminars.
>
> TB3: Seminars. And they said some students, because of the right brain, left brain thinking, can only understand when you write it down for them. Others must touch it, feel it, you know, like art. Draw it. And so, at that seminar, they said if you

can do both, the more you do, the more likely everyone — you'll catch somebody one way or another. So every time I went to a seminar, and I've been teaching for 29 years, I've gotten many, many, many tricks and things that I've learned that I do not get from the textbooks. Plus, you know what? That after you've taught a lot of times, when a student is like this, then you try something else, and if they're all going, then you think, OK. Then I'll move down. You know, they're shaking their head yes. They seem to understand, but when they go like this, and they're like bewildered.

Interview with TC1

TC1's classes observed were algebra with the textbook using UCSMP *Algebra* (McConnell et al., 1996), and precalculus/discrete mathematics using UCSMP *Precalculus and Discrete Mathematics* (Peressini et al., 1992).

The algebra class was to introduce "multiplying algebraic fractions", corresponding to Lesson 2-3 of the textbook (McConnell et al., 1996, pp. 85–90). Although the topic seemed relatively easy for students, TC1 especially drew their attention to the fact that the denominator of a fraction must not be zero. The conversation below reveals that TC1 considered it a key concept and students would have trouble with it when a denominator contains a variable.

> Interviewer: OK. How did you know you needed to stress to students that denominator of a fraction could not be zero?
>
> TC1: Because that's a key concept once you introduce the idea of a variable in the denominator. Because when it's numbers, there's no problem. If there's a zero in the denominator, most of the students will say, oh, that doesn't look right. They've seen that enough to know that there shouldn't be a zero in the denominator. But once you put a variable in there, you want to make sure they remember that rule about not being zero is still there and that they have to consider that variable when they think about it.
>
> Interviewer: How did you know that's important for students not to make the mistake?
>
> TC1: Well I was taught that was important. How did I know that they'd make that mistake? Because I've seen that mistake over and over.
>
> Interviewer: So from your own teaching experience or learning experience?
>
> TC1: Yes.
>
> Interviewer: Both of them?
>
> TC1: Probably both, but more from the teaching experience. You learn more by teaching than you do by taking classes, I think.

Findings of the Chicago Study (II):Pedagogical Content Knowledge

When TC1 explained how to find areas and volumes using multiplication in application problems, he reminded students not to forget the units after getting the result of the mathematical operation. During the interview, he stated that he learned that from his student teaching experience.

Interviewer: In the algebra class, you made sure students did not forget units.

TC1: And they always forget units.

Interviewer: How did you know this?

TC1: That's experience. When I student taught here, I taught geometry. And these were the kids that would be the next class after this algebra class. Every test that I gave them, they would forget to put the units down. And you could remind them over and over and over.

Interviewer: So that's your own experience?

TC1: Yes. That's an experience that I've had teaching.

Interviewer: Not told by any other sources or other persons?

TC1: No. Now, [the math chair][11] has also stressed it to me for the same thing. He said, make sure they're good on the units.

Noticing that it is not clear if TC1's teaching belief of this matter was strengthened by his inservice teaching experience or by his colleague, the math chair, I coded the source as preservice training, because it is clear that TC1 primarily formed such a belief from his student teaching, a component of preservice training.

The second class, precalculus/discrete mathematics, was devoted to the topic, "completing the squares with quadratics", a lesson added by the teachers after Lesson 2-2 (Peressini *et al.*, 1992, pp. 87–92) for the purpose of reinforcing the concept and preparing students for Lesson 2-3 of the textbook (pp. 93–99).

Instead of giving students the ready-to-use formula:

$$ax^2 + bx + c = a\left(x + \frac{b}{2a}\right)^2 + \frac{(4ac - b^2)}{4a}$$

---

[11] The actual name is not disclosed here for purposes of anonymity.

TC1 employed two concrete examples to show students how to complete the square for a quadratic expression. He explained his knowledge base for the approach and what the source is during the interview.

> Interviewer: In precalculus class, you used the two concrete examples to show students how to do that. For me, the question is why you did not introduce the formula directly like this one [see above]?
>
> TC1: Because if you show them the general formula, they're going to want to just memorize the general formula, and they'll memorize it wrong. I don't have that formula memorized myself. I don't memorize formulas. Now, I can derive that. And I will show them that probably today, where that comes from, because then I can connect this with the quadratic formula, which they've already seen.
>
> Interviewer: Well for me, I think it seems in your approach yesterday, you preferred to use the concrete examples that show how to do that.
>
> TC1: Yes. I want them to try it first.
>
> Interviewer: So how did you know that would work?
>
> TC1: Because that's how I learned. Most of that, because that's how I learned was through examples. If there was a formula when I was a kid, I would write the formula down, but I still would try to figure out how to get there first.
>
> Interviewer: So that's why you designed two examples to show students how to do that.
>
> TC1: Yes. Because if I throw a bunch of letters up there first, they get all flustered. But if I show them a couple of examples first, and then we go into the form, they hopefully will able to handle it.
>
> Interviewer: OK. I've got it.

Finally, the following discussion tells TC1's knowledge of what students likely encounter most difficulty with in the precalculus class he taught and how he came to know it.

> Interviewer: In the precalculus class I observed, you asked students to complete a square like this one $(x^2 + bx + \underline{0})$. I wonder if this is the most difficult part of the lesson.
>
> TC1: Yes. That was the difficult part. They had a difficult time with it.
>
> Interviewer: How did you know that this was the most difficult?
>
> TC1: How did I know that they would have difficulty with it?
>
> Interviewer: Yes. I would not know that.
>
> TC1: OK. I remember this concept from when I was in high school and I remember my other fellow students not being able to do it.

Interviewer: From your high-school experience as a student?

TC1: As a student. I mean, I remember this concept as a difficult concept in high school for a student. I remember as a student. It wasn't that difficult for me, but it was difficult for my fellow students. And a lot of times I would have to help them with that concept.

It is clear that the teacher's experience as a student learning the topic gave him the impression of what the difficult thing is. In another case, he claimed, "I still remember my first high-school math teacher, who died during the first quarter I had him. I only had him for nine weeks, but I still remember the things he did." Other teachers also pointed out during the interviews that they remember some things in their school days very clearly.

Interview with TC2

The interview with TC2 about her PCnK focuses on the second observed class, a precalculus/statistics class, in which the teacher introduced "step functions", corresponding to Lesson 2-4 in UCSMP *Functions, Statistics, and Trigonometry* (Rubenstein et al., 1992, pp. 99–104). The other observed class, which provided more opportunities to observe the teacher's knowledge of technology rather than PCnK, was an introduction of "using automatic grapher to graph functions", an optional in-class activity in Chapter 2 of the textbook, UCSMP *Advanced Algebra* (Senk et al., 1996).

Before giving the definition of the ceiling function, TC2 first presented a real-life problem — O'Hare Parking Garage Problem. She gave the following information:

> O'Hare Intermediate Parking
> Drive In: $ 4.00
> Per Hour: $ 3.00

The teacher told students she once parked her car there for four and a half hours, and was charged for five hours. Asking students why it was charged that way, TC2 introduced the concept of "rounding up", and led them to find more real situations of "rounding up" and "rounding down", e.g., for rounding up — phone bills, sales tax, postal rates, hotel stays, and for rounding down — paychecks, working overtime, etc. Then TC2 gradually gave students the formal definitions, the notations, the evaluation, and the graphs of the functions.

The questions in the interview about the above approach started with asking TC2, "How did you design that way?"

> TC2: I did that way because this is the way I think about it.
>
> Interviewer: Because people usually give the definition first and I am interested in how did you get to know this way would work?
>
> TC2: I didn't want to do it that way, because I don't think they understand it with definition first. I think that I used to teach it that way. So this in now, I guess from past experience of teaching. I used to teach with it with definition first, but I find that they don't have a clue what it means to say it's the smallest integer blah, blah. They get it all confused and they can't understand why this goes up and down and everything. That's why.
>
> Interviewer: This is from your past experience.
>
> TC2: Past teaching experience.
>
> Interviewer: How do you know the O'Hare Parking Garage?
>
> TC2: I know the O'Hare Parking Garage because I used to park there almost every day. When I worked for [a company name][12] I would travel on the plane almost every day, so I could quote you the rates, I guess.

Also notably different from the textbook, the way TC2 used to represent "the step functions" was first to introduce the ceiling function,

$$f(x) = \lceil x \rceil$$

then introduce the greatest integer function (the floor function) $y = \lfloor x \rfloor$. When asked why she designed a different way to approach the topic, she explained:

> TC2: I purposely chose to do the step function lesson differently because I felt that why should we introduce the greatest integer function when that is not the one that has the application. It is much easier for a student to make reference to the other one, the ceiling function than it is the floor function, and so as a result I started with the ceiling function, even though mathematically most books will start with the floor function. But you can see it was much easier for them to get a frame of reference from the parking lots.
>
> Interviewer: How do you know it's much easier?

---

[12]The actual name is not disclosed here for anonymity.

TC2: I think it's going by the premise that if they can relate to it in any circumstance that they have encountered they will then dig into their brain memory to recall that, and so to make a connection. They'll make a brain connection. How did I learn that?

Interviewer: How did you get this knowledge?

TC2: I got that because I went to a seminar in brain studies in middle of August of this year. And one of the things that they were talking about at the seminar was how always the brain makes connections and so we had been saying that in mathematics a lot, but they were stressing that you really do have to make reference to something that is known prior. And so I was using basically, I think, the premise. What is known prior? Well, they may not know the notation, so I said they really know this thing by experiencing it and what is the experience? The ceiling function. Most of the time you round up in real life, and therefore I counted that as the connection.

Because graphing calculators such as TI-82 and TI-83 can only draw the floor function $y = \lfloor x \rfloor$, but not the ceiling function $f(x) = \lceil x \rceil$, the teacher presented a special way to represent the ceiling function, that is, $y = -\lfloor -x \rfloor$ to graph the ceiling function using the graphing calculator. When asked how she got to know that, TC2 pointed out,

I knew that because I could not figure it out on the calculator and a colleague, I happened to ask a colleague, and the colleague said to me, oh, I think all you have to do is this, this is what I did and that's how I found out. A colleague told me.

Note the above special way to represent the floor function is given in the teachers' edition of the textbook (Rubenstein *et al.*, 1992, p. 100). It indicates that teachers sometimes do not fully use textbooks as a source for their PCnK, and exchanges with colleagues can be a more convenient source. As a matter of fact, it is also a more important source than textbooks as revealed from the questionnaire data discussed in the previous section.

Interview with TC3

TC3's first class I observed was a geometry class, covering the topic: "drawing conclusions" in the textbook, *Geometry: For Enjoyment and Challenge* (Rhoad *et al.*, 1991, Lesson 2-4, pp. 72–75).

The interview about the PCnK began with the fact that, to approach the topic, the whole class period was devoted to students' working and discussing the questions in Problems Set A (1–7) and Set B (8–12) in the textbook.

> Interviewer: In your geometry class, you used the twelve questions in the textbook. And you asked students to give their answers and then you explained it. I'm unsure why you decided to use all of the twelve questions, instead of just selecting them.
>
> TC3: Because this was such a strong foundation, I found in the past if I didn't take the time to go through every single one of those questions, they would have great difficulty. In other words, they're not all the same. Each different question that they put up had a very different concept.

The interview also revealed TC3's knowledge of what is most important or difficult for the geometry class, as she stated:

> I know that it's very important that they understand the chain of reasoning, that they understand the logic that they'll have to — they have very much difficulty between a conditional and its converse. So it's very important in the beginning of the geometry course to make sure they understand that if they're going to start with "if p", somewhere in the proof above that, they have to have said that "p" is true. (Interviewer: How did you get to know that?) Because I've taught this course for five or ten years. I've seen the mistakes in the students' reasoning. And so, from the past experience, I know what the pitfalls are. I know where they're going to have trouble. So I know what to emphasize. And I know that they have trouble knowing whether it's "if bisect, then congruent angles" or when they do "if congruent angles, then bisect". So I know to take the time to do very short draw conclusions.

The second class observed was algebra for underperforming students, using the teaching materials developed by the mathematics department of the school. The class period was for introducing the concept of "the unit circle", in which the sub-concept of "the reference angle" (see the figure below) is particularly difficult for students, and TC3 used many graphical representations to show students the concept and solve problems related to the concept.

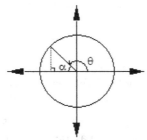

α is the reference angle formed by the terminal side of θ and the x-axis.

Findings of the Chicago Study (II):Pedagogical Content Knowledge 151

The excerpt below explains how TC3 developed her knowledge that using graphical representation would be helpful to teach the class.

> Interviewer: In algebra class, you used a lot of graphical representation.
>
> TC3: These students are visual learners. They need to see things. They have a great deal of difficulty with anything abstract. You have to be very specific, very concrete. You cannot talk in abstract terms, because their minds are not that developed.
>
> Interviewer: Yes. Now my question is, how did you or when did you begin to be aware it's easy for students to learn visually or use the graphical representation?
>
> TC3: Well actually, three things. One thing would be my past experience. The second thing would be working with my colleagues and seeing them try visual things. And the third thing is I attended a lot of conferences.

In concluding this section, Table 6.6 provides a summary of the interview data analyzed above, showing for each source listed, how many teachers mentioned that it has contributed to the development of their

Table 6.6 Sources of teachers' PCnK (from interview data)

Source[a]	A	B	C	D	E	F	G	H
TA1					yes (3)			
TB1	yes (2)						yes (2)	
TC1	yes (3)	yes (1)					yes (1)	
TA2					yes (1)		yes (3)	yes (1)
TB2	yes (1)						yes (2)	yes (1)
TC2			yes (1)		yes (1)		yes (1)	
TA3							yes (4)	
TB3			yes (1)				yes (3)	yes (1)
TC3				yes (1)	yes (1)		yes (3)	
Total[b]	3 (6)	1 (1)	2 (2)	1 (1)	3 (3)	0	9 (22)	3 (3)

[a] A = Experience as student, B = Preservice training, C = Inservice training, D = Organized professional activities, E = Informal exchanges with colleagues, F = Reading professional journals and books, G = Own teaching experience and reflection, H = Textbooks.
[b] The figures in parentheses are the numbers of different kinds of PCnK that teachers explained that they learned (regardless how much) from the corresponding sources.

PCnK of the topics taught in the classes observed, and for each teacher interviewed, how many sources by which he/she obtained his/her PCnK used in the classes observed.

From the table and the discussion earlier, we have no doubt that teachers' own teaching experience and reflection is the most frequently used source in developing their PCnK. From their teaching experience, teachers learned and accumulated a wide range of PCnK about mathematics topics taught in their classes, such as what is important and should be emphasized (e.g., TA1, TC3), what is difficult (e.g., TC1, TA3, TB2, TC3), what is not difficult (e.g., TA2, TB2), what way works to represent a concept (e.g., TC2, TB3), what does not work (e.g., TC2), what technique helps students to avoid mistakes (e.g., TB2), what mistakes students will easily make (e.g., TA2, TB3), and so on.

It is also noticeable that, according to the interview data, no teachers described "reading professional journals and books" as a source of their PCnK showed in the classes observed. Although somewhat unexpected, I think this result clearly reflects that it is by no means an important source of teachers' PCnK. The reason I believe is that teachers on the one hand did not often read the professional journals and books (see more details in Chapter 8), and on the other hand, not many such professional references are available to teachers.

Last, the interview also provided empirical evidence that teachers learned their various PCnK from many other sources. In addition to their own teaching experience and reflection, they also developed their PCnK from their experience as students, inservice training, daily exchanges with colleagues, and so forth. It is hard not to compare the relative importance of those sources, as well as it not being the purpose of the interviews.

Summary of the Findings

This chapter is devoted to the findings of how teachers developed their PCnK. The issue is investigated through the questionnaire survey, which contained two open-ended questions to ask teachers what topics they recently taught, how they represented them to students, and how they got to know the ways to represent; through the classroom observations with nine teachers which were designed to identify what PCnK those teachers have about the topics taught in the classes; and through the interviews with the nine teachers, which were intended to gain empirical evidence and a detailed look at how teachers actually developed their PCnK demonstrated in the classes observed.

Three main findings can be drawn from the statistical analysis of the questionnaire data and the case-by-case examination of the interview data.

First, there are various sources from which teachers develop their PCnK that they use most often in their daily teaching activities.

Second, relatively speaking, the most important sources for teachers to develop their PCnK are their "own teaching experience and reflection" and their "informal exchanges with colleagues". The secondarily important sources are "textbooks", "experience as student", "organized professional activities", "inservice training", and "reading professional journals and books". The least important source is "preservice training". Statistically, setting "organized professional activities" as a criterion, the sources in the first group are significantly more important at the 0.05 level, the sources in the second groups have no significant differences in terms of importance, and the source in the third group, namely, "preservice training", is significantly less important at the 0.05 level.

Furthermore, all of those sources have the same importance to teachers with different amounts of teaching experience, except for the source of "informal exchanges with colleagues", from which younger teachers learned significantly more than their senior colleagues.

Chapter 7

Findings of the Chicago Study (III): Pedagogical Instructional Knowledge

This chapter is intended to address the question of how teachers develop their knowledge of teaching strategies and techniques, namely, their pedagogical instructional knowledge (PIK). As discussed in Chapter 3, teachers' PIK is defined in this study, based on the NCTM *Standards*, as their knowledge of instructional strategies and classroom organizational models. In some sense, a more conventional synonym of it used by researchers is "general pedagogical knowledge", which according to Shulman means "those broad principles and strategies of classroom management and organization that appear to *transcend* subject matter" (Shulman, 1987, emphasis added), or sometimes "pedagogical knowledge" itself, when it is narrowly confined to "teachers' knowledge of general teaching procedures such as effective strategies for planning, classroom routines, behavior management techniques, classroom organizational procedures, and motivational techniques" (Fennema & Franke, 1992, p. 162). In short, the instructional strategies and classroom organizational techniques of PIK are about teaching, but usually not particularly limited to teaching mathematics, and absolutely not limited to teaching a special mathematics topic (this is one important difference from PCnK, discussed in Chapter 6).

Also, I want to make a distinction in this study between teachers using teaching strategies (or techniques, or methods, or models, or skills) and teachers' other teaching behaviors. Namely, teachers' strategies are used steadily and are often pre-determined; in other words, their teaching

156   *Investigating the Pedagogy of Mathematics*

strategies are internalized and are likely (though not definitely so) to be used by themselves under certain teaching conditions. In contrast, teachers' other teaching behaviors are often their instant or natural reaction to an uncertain teaching environment, and do not have a steady pattern. For example, if a teacher does not review previous work at the beginning of a class period just because he feels there is not enough time in this period, then it is not considered to be a strategy in this study. In fact, the teacher might start with reviewing in the next period when there is time (starting with reviewing is actually his strategy). However, if he usually does not review previous work because he realizes that makes students assume they can always wait for another chance to catch up, then not reviewing at the beginning of a class is considered to be his teaching strategy. The knowledge bases for those two kinds of teaching behaviors may be different.

There have been countless theoretical or practical books, e.g., Borich (1992), Good & Brophy (1994), and Myers & Myers (1995), to name but a few (let alone journal articles), devoted to investigating and/or disseminating effective teaching strategies and classroom management models, ranging from general methods such as how to use cooperative learning to specific techniques such as how to ask questions and how to use wait-time in classroom teaching.

In this study, to determine how teachers develop their PIK, a questionnaire survey, classroom observation, and interviews following the classroom observation were used. Reported below are the findings of the study, based on the analysis of the data collected from those data sources, respectively.

Analysis of the Questionnaire Data

Question 19 of the questionnaire (see Appendix 1A) is about how teachers develop their PIK. All the 69 teachers answered this question. Table 7.1 summarizes their responses by displaying the distributions of the numbers of teachers giving different evaluations of the various sources.

From the table, it is clear that there are a variety of sources from which teachers developed their PIK, and for different teachers, their main sources for developing their PIK could be different. More importantly, in terms of the combined percentages of teachers' choosing positive evaluation of

Findings of the Chicago Study (III): Pedagogical Instructional Knowledge

Table 7.1 Distributions of the numbers of teachers giving different evaluations about the contributions of various sources to their PIK

Sources	Degree of the contribution			
	Very much	Somewhat	Little	No contribution
Experience as student	6 (8.7%)	22 (31.9%)	21 (30.4%)	20 (29.0%)
Preservice training	5 (7.2%)	22 (31.9%)	29 (42.0%)	13 (18.8%)
Inservice training	21 (30.4%)	31 (44.9%)	11 (15.9%)	6 (8.7%)
Organized professional activities	12 (17.4%)	29 (42.0%)	21 (30.4%)	7 (10.1%)
Informal exchanges with colleagues	50 (72.5%)	17 (24.6%)	2 (2.9%)	0 (0.0%)
Reading professional journals and books	(5.8%)	15 (21.7%)	34 (49.3%)	16 (23.2%)
Own teaching practices and reflection	60 (87.0%)	8 (11.6%)	1 (1.4%)	0 (0.0%)

Note: n = 69. The figures in parentheses are percentages of teachers giving the corresponding evaluation. The sum of the percentages in each row might not be exactly 100% due to rounding.

"very much" and "somewhat", overall teachers' "own teaching practices and reflection" (98.6%) and "informal exchanges with colleagues" (97.1%) are the two most important sources for teachers to gain their PIK; the contributions of "inservice training" (75.4%) and "organized professional activities" (59.4%) seem also relatively highly recognized; while "experience as student" (40.6%), "preservice training" (39.1%), and "reading professional journals and books" (27.5%) are the least important sources.

Based on teachers' average evaluation in the question, Figure 7.1 presents a general comparison of the contribution of different sources to teachers' PIK.

We can see that, in terms of teachers' average evaluation shown in Figure 7.1, the order of the sources for teachers to develop their PIK is, from the most important to the least important, "own teaching experience (practices)

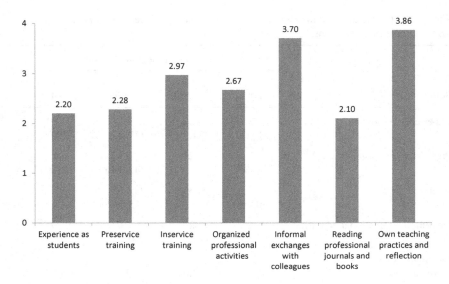

Figure 7.1 Comparison of the contribution of different sources to teachers' PIK

Note: By the ordinal scale in the figure, 4 = very much, 3 = somewhat, 2 = little, and 1 = no contribution.

and reflection" (3.86), "informal exchanges with colleagues" (3.70), "inservice training" (2.97), "organized professional activities" (2.67), "preservice training" (2.28), "experience as student" (2.20), and "reading professional journals and books" (2.10), consistent with the order of percentages with positive evaluation in Table 7.1, except for "preservice training" and "experience as student".

Based on the above preliminary analysis, log-linear regression analysis was further applied to the data. Represented in Table 1F.4 in Appendix 1F is the main result of "PROC LOGISTIC" procedure using SAS, which includes the maximum likelihood estimates of the parameters and the corresponding significant tests.

Following the same principles used with the log-linear regression analysis in Chapter 5, we know that, according to the parameter estimates in Table 1F.4, the order of importance of the sources to the development of teachers' PIK are "own teaching practices and reflection" (G: −3.4997), "informal exchanges with colleagues" (E: −2.5913), "inservice training"

Findings of the Chicago Study (III): Pedagogical Instructional Knowledge 159

(C: −0.7146), "organized professional activities" (D: 0), "preservice training" (B: 0.8041), "experience as student" (A: 0.9635), and "reading professional journals and books" (F: 1.1670).

It should be noted that the order of the importance of the different sources based on the above logistic models are the same as revealed by teachers' average ratings of the contribution of the sources, shown in Figure 7.1 above.

Moreover, from the values of Pr > Chi-Square in Table 1F.4, we see that, compared to the source of "organized professional activities", teachers' "own teaching experiences and reflection", "informal exchanges with colleagues", and "inservice training" are significantly more important at the 0.05 level, while "preservice training", "experience as student", and "reading professional journals and books" are significantly less important.

Regarding how the length of teachers' teaching experience is related to their evaluation of the contribution of different sources to their PIK, again, I classified all the teachers into three groups: TG1 consisting of teachers with 0–5 years' teaching experience, TG2 with 6–15 years', and TG3 with 16 or more years'.

Figure 7.2 presents the average evaluation of the teachers in those three groups of the contribution of different sources to the development of their PIK, from which we can see how their overall evaluation differs across those three groups.

The figure shows that the biggest differences exist for Source B ("preservice training"), with the difference between TG1 and TG2 being 0.65, and for Source D ("organized professional activities"), with the difference between TG1 and TG2 being 0.64; all other differences are within 0 to 0.41. Overall, the differences of the average evaluations among those three groups for all the sources seem relatively small.

Table 7.2 shows the actual distributions of the numbers of the three groups of teachers giving different evaluations of the contribution of each source to their PIK. The chi-square test for each source is also included in the table so we can see if there is a statistically significant difference among those three groups.

At the 0.05 level, the p-values of the chi-square test in Table 7.2 reveal that there is no significant difference among the distributions across those three groups for each source. Combining the result with the descriptive

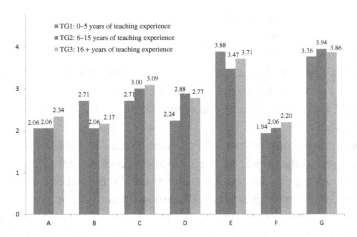

Figure 7.2 Comparison of the three groups of teachers' average evaluations of the contribution of different sources to their PIK

Note: 1. For sources, A = Experience as student, B = Preservice training, C = Inservice training, D = Organized professional activities, E = Informal exchanges with colleagues, F = Reading professional journals and books, G = Own teaching practices and reflection.
2. The evaluations are shown by the ordinal scale in the figure, 4 = very much, 3 = somewhat, 2 = little, and 1 = no contribution.

comparison shown in Figure 7.2, we can see that the length of teachers' teaching experience did not play a significant role in teachers' viewing of the importance (contribution) of different sources to the development of their knowledge of general instructional strategies and classroom management skills. In other words, each of the sources carries basically the same importance to teachers with different lengths of teaching experiences in developing their PIK.

Analysis of the Interview Data

As mentioned previously, the interviews of this study with nine teachers were conducted after the observation of their classes, which were expected to provide specific contexts for the researcher to detect what knowledge teachers have through observing their classroom teaching behavior. The underlying reason for this research method is that any kind of a teacher's teaching behavior, from adopting general strategies to using specific techniques, is essentially the result of his/her decision-making,

Table 7.2 Distributions of the numbers of teachers among the three groups giving different evaluations of the contribution of each source to their PIK

Source A: Experience as student

Evaluation	TG1(0–5 years)	TG2(6–15 years)	TG3(16+ years)	Total
Very much	0	1	5	6
Somewhat	6	5	11	22
Little	6	5	10	21
No contribution	5	6	9	20
Total	17	17	35	69

Chi-square test: $\chi^2 = 3.5290$ df $= 6$ p $= 0.7401$

Source B: Preservice training

Evaluation	TG1(0–5 years)	TG2(6–15 years)	TG3(16+ years)	Total
Very much	2	1	2	5
Somewhat	10	3	9	22
Little	3	9	17	29
No contribution	2	4	7	13
Total	17	17	35	69

Chi-square test: $\chi^2 = 9.9218$ df $= 6$ p $= 0.1280$

Source C: Inservice training

Evaluation	TG1(0–5 years)	TG2(6–15 years)	TG3(16+ years)	Total
Very much	3	4	14	21
Somewhat	9	10	12	31
Little	2	2	7	11
No contribution	3	1	2	6
Total	17	17	35	69

Chi-square test: $\chi^2 = 6.9005$ df $= 6$ p $= 0.3301$

(*Continued*)

Table 7.2 (*Continued*)

Source D: Organized professional activities

Evaluation	TG1(0–5 years)	TG2(6–15 years)	TG3(16+ years)	Total
Very much	0	5	7	12
Somewhat	7	6	16	29
Little	7	5	9	21
No contribution	3	1	3	7
Total	17	17	35	69

Chi-square test: $\chi^2 = 7.0450$ df $= 6$ p $= 0.3167$

Source E: Informal exchanges with colleagues

Evaluation	TG1(0–5 years)	TG2(6–15 years)	TG3(16+ years)	Total
Very much	15	10	25	50
Somewhat	2	5	10	17
Little	0	2	0	2
No contribution	0	0	0	0
Total	17	17	35	69

Chi-square test*: $\chi^2 = 8.6634$ df $= 4$ p $= 0.0701$

Source F: Reading professional journals and books

Evaluation	TG1(0–5 years)	TG2(6–15 years)	TG3(16+ years)	Total
Very much	0	1	3	4
Somewhat	3	4	8	15
Little	10	7	17	34
No contribution	4	5	7	16
Total	17	17	35	69

Chi-square test: $\chi^2 = 2.6091$ df $= 6$ p $= 0.8561$

(*Continued*)

Table 7.2 (*Continued*)

Evaluation	Source G: Own teaching practices and reflection			
	TG1(0–5 years)	TG2(6–15 years)	TG3(16+ years)	Total
Very much	13	16	31	60
Somewhat	4	1	3	8
Little	0	0	1	1
No contribution	0	0	0	0
Total	17	17	35	69
Chi-square test**: $\chi^2 = 2.4971$ df $= 2$ p $= 0.2869$				

*Because here all the cells in the last row are 0, the chi-square test cannot be directly applied. The results were obtained by applying the test to the rest of the table, excluding the last row.

**The results were obtained by applying the test to a 3 × 2 table, which consists of the 1st row and a second row combining the original 2nd and 3rd rows to meet the common assumption of using the chi-square test. The original last row was excluded as all the cells in the row equal 0.

which in turn rests on the knowledge he/she possesses (Fennema & Franke, 1992).

The classroom observation revealed that teachers used a wide range of teaching strategies and techniques in their teaching. Some are commonly used and well recommended in recent years, such as cooperative learning; some are more subtle and only used by a few teachers, such as asking students voluntarily to give answers; and some are very special and only used by one or two teachers, such as having students present orally in front of the whole class.

After the classroom observation, each teacher's four to six teaching strategies and techniques, general such as cooperative learning or special such as use of an overhead, demonstrated in the classrooms were clearly (though naturally not exhaustively) identified, then questions about how the teacher developed those kinds of instructional knowledge were raised during the interviews with that teacher. A few general teaching strategies, such as "asking students to do reading and writing in mathematics", also received attention during the interviews even though they were not necessarily used in some teachers' classes observed.

Described below is an analysis of the collected data. For each teacher observed and interviewed, I first utilize some detailed excerpts from the conversation to provide specific insight into how each teacher developed his/her knowledge of some (usually two or three) particular teaching strategies and techniques from different sources. Following that, I shall use a table, which is based on the coding system of this study, to summarize what was found for each teacher about the teaching strategies and techniques identified and the knowledge sources explained by the teacher.

It should be pointed out that the identified list of each teacher's teaching strategies and techniques is by no means exhaustive. The main reason is that I only observed two classes for each teacher and it is unlikely that teachers would use all teaching strategies or techniques that they know in just two classes. In addition, the list did not include a teacher's teaching strategies or techniques which he/she did not remember or explain what the main knowledge sources were. For example, TA3 said during the interview that she did ask students to "do writing" in mathematics, but she did not explain how she developed such an idea. Another example is, when asked how she got her knowledge of using the overhead for teaching mathematics, TB3 responded: "there is nothing to learn [with the overhead]". By the way, I also did not include "using technology" and "using concrete materials" in this chapter as they have been considered in Chapter 5 based on the conceptual framework of this study, though they can, in some sense, be thought to be "teaching strategies".

Interview with TA1

Anyone who for the first time visited the school at which TA1 was working would be impressed that all mathematics classes use cooperative learning; students are organized to sit together in groups of four in classrooms all the time and work cooperatively. Actually, all mathematics teachers, regardless of young or old, are expected and virtually required to use the strategy. However, did all teachers develop their knowledge of cooperative learning the same way? Here is TA1's explanation about how she developed her knowledge of that strategy.

> Interviewer: Regarding general teaching styles, I noticed that you used a lot of cooperative learning [in the classes I observed]. Like in geometry honors class,

you distributed two sheets of paper and asked students to work in groups and then get them to switch and check answers with each other. Why did you do that?

TA1: I do things like that a lot with them. I'm a firm believer that, I notice with my teaching experience, that if I can explain to you how to do something, [or] I can show you how to do something, then that means I really understand how to do it. So making them verbally explain it to somebody else, and just helping each other out. I think one of the most valuable resources they have are [is] each other. And I think I found that out from [my own experience] — I don't think anybody told me that.

Interviewer: When did you first time know the strategy of cooperative learning?

TA1: I was introduced to it a little bit in college, and then student teaching we used it. We used it more informally. Here, their desks are in groups.[1] When I student taught, it was more [like that] let's do this as a group activity, and they were still in rows; they had to move together in their groups, and they didn't have necessarily set jobs that they had to do. Where I think now, it makes it — I have much more instruction on that they know what they need to do and what they need to complete.

Another teaching strategy identified with TA1 was using warm-up questions at the beginning of both the observed classes. During the interview, she told me that she always did that. When asked about the source of that knowledge, she mentioned, in addition to her preservice training, how a professional book particularly helped her in this aspect.

Interviewer: Why do you always have a warm-up question?

TA1: Because it gets them [students] in the room. It gets them with suppliers out and they are already thinking when the bell rings.

Interviewer: How did you get to know that?

TA1: I first started that when I was student teaching. My co-ops used the idea a little bit. I'd also read about it in an education book somewhere. And then, of course working here with students and faculty, I think everyone just about has [used the same technique].

Interviewer: You mean an education book you read here?

TA1: I got it from — this is interesting. Actually, I read it in college in probably one of my most valuable education classes that I had in my undergraduate [work]. But the book that I read there — when I came here to start teaching, [The math chair][2] gave me the same book. And I said, I've already read it. So it was something that

---

[1] From the classes I observed and what I was told by the math chair, this is the case for all mathematics classes in the school.

[2] The actual name is not disclosed here for purposes of anonymity.

I think kind of neat. There was so much in that book that I did believe in. So it was kind of nice coming and going, oh, well, these are the same beliefs.

Interviewer: Is that education book a textbook?

TA1: No. It's a book called *Making Minutes Count*,[3] and then the author has a second one called *Making Minutes Count More*,[4] and it's almost — it's a little book, about this thick. It's more — it's book from somebody that taught for many years. And it was a little, structural thing that he thought would be very useful. And he passed them on.

TA1's explanation about how her experience as a student helped her to realize the importance of using the overhead for classroom teaching revealed that a teacher can learn not just from what they liked, but also what they didn't like, from the way their teachers taught them. She said:

I feel very uncomfortable when I'm up at the board and I constantly have to write something and them look and write, because I want to be facing them and I want to see them and I want to see their reactions. (Interviewer: How do you know that?) I think that was experience [as a student]. When I was a student, when my teacher was rushing or had bad handwriting, and you're trying to copy and you go, what is that word? What does that mean?

The other two sources she mentioned are preservice training (her co-op teachers told her to use it) and her own teaching experience and reflection.

Table 7.3 presents the general teaching strategies (and techniques) identified with TA1 and the knowledge sources she pointed out during the interview.

Interview with TA2

Although also using cooperative learning, TA2 believed that his colleague (the math chair) was the only main source for him to use that strategy, and he explicitly excluded other things such as attending conferences and reading professional journals and books as his knowledge source.

Interviewer: About the teaching strategy, you used cooperative learning in the classes. When did you get to know [this strategy]?

---

[3] The actual name of the book is *Every Minute Counts*, written by David R. Johnson (1982). In Chapter 3 of the book: "The new class routine", the author discussed how to effectively use the first five minutes of a mathematics class, to which TA1 referred here.

[4] The actual name is *Making Minutes Count Even More* (Johnson, 1986).

Table 7.3 Teaching strategies and the knowledge sources (TA1)

Teaching strategies	Sources[5] identified
1. Cooperative learning	B: preservice training; G: own teaching experience and reflection
2. Ask student to do writing	B: preservice training; G: own teaching experience and reflection
3. Always have warm-up questions	B: preservice training; F: reading professional books
4. Clearly tell students the lesson objectives at the beginning of the class	A: experience as student; B: preservice training; E: exchanges with colleagues
5. Use the overhead and uncover the transparency gradually	A: experience as student; B: preservice training; G: own teaching experience and reflection

TA2: That is something [the math chair] has actually insisted us all use. I doubt I would have done it on my own. [The math chair] actually forced us to try that, and it worked OK.

Interviewer: But is [the math chair] the only source for you to do that?

TA2: He's the only source that I've used. There are a lot of conferences and talks available about it, but I have not gone to any.

Interviewer: Have you read some articles in magazines about the idea?

TA2: I skipped those articles.

The interview also revealed that TA2 rarely required students to read mathematics in his teaching. However, he did ask students to do writing. It is interesting to see how he developed such a teaching strategy.

Interviewer: Did you use writing or reading? I mean, ask them [students] to write or read in your classroom or as homework?

TA2: In Algebra C, they had journal assignments that they write once per week, where I give them a math question and they have to write about it.

---

[5]Note: for Tables 7.3–7.12, A = Experience as student, B = Preservice training, C = Inservice training, D = Organized professional activities, E = Informal exchanges with colleagues, F = Reading professional journals and books, G = Own teaching experience (practices) and reflection.

Interviewer: OK. When did you start to do that in your teaching?

TA2: I've been doing that, I think, since I started teaching [ten years ago].

Interviewer: Yes. So, how did you know that?

TA2: It's something I tried because some of my colleagues in [previous school taught] were doing it. I did it because I was trying to get to know my students better initially. I started out with topics that were not math related. They were like school related. So that I could develop a rapport with them. Then I realized that I could at least put a little math in it and get an idea of do they understand the concept order? Do they need help with the concept further? So, it slowly evolved into putting math into it.

Interviewer: Then you learned from your own experience how to do that or did you get the idea from other sources?

TA2: [From my experience.] Well, I read articles also about teachers who had also actually done all their journals about only math. And so then, I started to do only a little bit of math and started mixing. And now I do mostly math-type stuff, rather than personal journals. So really, the idea to start changing came from articles that I was reading.

Table 7.4 summarizes the teaching strategies identified with teacher TA2 and the knowledge sources he described during the interview.

Interview with TA3

The interview with TA3 revealed there are many personal teaching strategies and techniques that teachers can learn from their own learning and teaching experience and reflection.

Table 7.4 Teaching strategies and the knowledge sources (TA2)

Teaching strategies	Sources identified
1. Cooperative learning	E: exchanges with colleagues[a]
2. Ask student to do writing	E: exchanges with colleagues; F: reading professional journals; G: own teaching experience and reflection
3. Clearly tell students the lesson objectives at the beginning of the class	B: preservice training; E: exchanges with colleagues
4. Selectively assign homework	G: own teaching experience and reflection

[a]Here and below, the math chairs are considered as teachers' colleagues.

Findings of the Chicago Study (III): Pedagogical Instructional Knowledge 169

It is interesting to note TA3's explanation about how she generated the idea of using cooperative learning for teaching mathematics from her experience as a student but in science courses, not in mathematics courses.

> Interviewer: I noticed that you used a lot of cooperative learning [in your classrooms]. Did you use it before, like ten years ago, or five years ago?
>
> TA3: I have been using cooperative learning for 20 years. I've used cooperative learning long before it became popular. I've been doing that for 20 years.
>
> Interviewer: OK. So now why do you know or how did you get to know such kind of teaching strategies?
>
> TA3: Why did I do it 20 years ago before everybody else was doing it? I guess....
>
> Interviewer: I mean what's the knowledge base to do that?
>
> TA3: I guess when I came out of college, I taught my classes the way I wished my [math] teachers had taught me.
>
> Interviewer: You mean college professors?
>
> TA3: No. Well, any [math] teachers.... Probably science and English teachers had more effect on me than my math teachers. Science teachers have labs. And labs are where students get together to work together and do experiments and do write-ups. I do that as my cooperative learning. I liked science when I was in school.

During the interview, TA3 also told me that she always prepared all the transparencies for overhead use ahead of time, and this was not the case five to eight years ago. She explained the reason for the change is for efficiency. When asked how she got to know that and if someone told her that, she answered:

> It was [my own] experience. I would make one up. And then I would realize that went fast. I'll do more like that tomorrow.

The teaching strategies identified with TA3 and the knowledge sources she explained are generally displayed in Table 7.5.

Table 7.5 Teaching strategies and the knowledge sources (TA3)

Teaching strategies	Sources identified
1. Cooperative learning	A: experience as student; G: own teaching experience and reflection
2. Use different colors on the blackboard	G: own teaching experience and reflection
3. Use pictures to visually represent ideas	A: experience as student; G: own teaching experience and reflection
4. Prepare all transparencies for the overhead use ahead of time	G: own teaching experience and reflection

Interview with TB1

In the interview, when asked what the general principle guiding his teaching styles was, TB1 first explained to me that when introducing a concept, he likes to let students work by themselves first.

> Interviewer: Now I would like to ask you some teaching strategies. From the two classes I observed, you used review, lectures, demonstration, examples, student independent work, a lot of whole class discussion. What's the general principle guiding your teaching styles, or strategies?
>
> TB1: Right. When I first introduce a concept, I like to have the students work by themselves, so they can — and this is something I tell them. They have to decide how well they understand the concept. Now if I go back to that concept the second day, I'll allow them to work in groups, because now either the student understands the concept or maybe they're struggling with the concept. And if they can hear it maybe from a different person, the concept, then maybe they might understand it a little bit better. So basically, what I'd like to do is first introduce a concept, have the students work by themselves.
>
> Interviewer: Yes. And how do you know that's important? That it will help them?
>
> TB1: Because that's just my own philosophy. That's how I was taught. You have to take it upon yourself to understand the concept. Then just over the course of the years, and talking to other teachers around here, I've come to understand the fact that maybe they don't understand the way I explain it, but maybe a friend of theirs might be able to explain it in a different manner. So that's why I don't mind putting them in groups.

In TB1's geometry class, one student was asked to orally present her work in front of the board to the whole class. That was not often seen in all the 18 classes observed in this study. When asked why he had such an idea for teaching, he insisted he learned that from his own teaching experience.

> Interviewer: Also I noticed in the geometry class, a female student was asked to present her work to the whole class. ... Because that to me is unusual when I observed other classes. I mean that kind of student presentation. Is that what you did often?
>
> TB1: Oh, yes. Sure. I love doing that to students. You know, having them go up to the overhead or chalkboard to present problems.
>
> Interviewer: So how did you get to know it's a good strategy? I mean, it works for students.
>
> TB1: I think it just goes back to my first year of teaching. I was always open to try new things. And I noticed that if a student was at the board, either they knew exactly what they were doing, or if they did make a mistake, they were able to see it as they were doing it. They're like, oh, maybe that doesn't make sense. And they'd correct it themselves.
>
> Interviewer: Well for me, I got such knowledge from you. Before you, how did you get this kind of knowledge? I mean, how did you — when did you have such an idea?
>
> TB1: I think it was just my first year of teaching, my teaching experience.

In his classes observed, TB1 not only wrote down his explanations and answers to examples very carefully and completely, but also required students to do so in their work. During the interview, TB1 also described how his own learning (in both formal and informal environments) and teaching experience played a role in his developing such a philosophy. He definitely excluded other sources.

> Interviewer: I noticed that you emphasize writing a lot [in the classes]. Maybe it's about — I mean, kind of communication. For example, in algebra class, you asked them to do the homework neatly. Why did you emphasize that? That's very interesting to me. I mean, you obviously want students to be good communicators and they write good. But why — I mean now it's stressed in many professional documents and journals. So when did you begin to notice that?
>
> TB1: I think it goes back to the teaching experience again.
>
> Interviewer: Here?
>
> TB1: Here, correct. And the only place I've really taught is here. I think this is a personal belief and it came about my first year. I always thought students should take pride in their work. Do the little things right.
>
> Interviewer: Yes. But many teachers do not ask them to write so much.

TB1: I think it's important that they understand the answer.

Interviewer: Where did you get that belief?

TB1: I think it was the way I was probably brought up by my parents and the structure I had with the education that I did have. It was always like that. There was no ifs, ands or buts, you know. You do things properly and fully and completely.

Interviewer: Your own life experience?

TB1: Yes. I'd say life experience, with the parents and everything.

Interviewer: Were you influenced by reading NCTM documents?

TB1: No. No.

Interviewer: Journals or magazines?

TB1: No.

Interviewer: And from your colleagues?

TB1: No. It's just me.

Overall, the teaching strategies identified with TB1 and the knowledge sources he explained are given in Table 7.6.

Interview with TB2

The conversation with TB2 about cooperative learning again proves that teachers can develop their pedagogical knowledge in different ways, some seeming very occasional.

Interviewer: [In the classes I observed] you used a lot of time on cooperative learning. I would like to know when was the first time for you to know cooperative learning?

Table 7.6 Teaching strategies and the knowledge sources (TB1)

Teaching strategies	Sources identified
1. Let students work on a concept first by themselves, then work in groups	A: experience as student; E: exchanges with colleagues
2. Ask students to orally present in front of the board to the whole class	G: own teaching experience and reflection
3. Teach students to write on mathematics completely and nicely	G: own teaching experience and reflection; Other: life experience
4. Encourage students to guess and check when doing problems	G: own teaching experience and reflection
5. Ask students to orally present a concept	E: exchanges with colleagues

Findings of the Chicago Study (III): Pedagogical Instructional Knowledge

TB2: I was doing student teaching in college. [About seven years ago] I was student teaching and one of the teachers at the high school that I was student teaching at was going to a presentation on cooperative learning. From, coincidentally, from the man who used to be the head of the math department here and in fact the man who hired me here, [name].[6] And he gave the presentation on cooperative learning and I was extremely impressed. Ever since then I've been really hooked on it.

Interviewer: So, this is the source of your knowledge?

TB2: Yes. ... And a lot of it comes from my own experience as a teacher. When I have my students in rows I feel they're not picking up as much as when they're actually working on something together.

Doing projects is a relatively innovative idea in mathematics teaching (for a brief introduction of "project-based learning", see Good & Brophy, 1994, pp. 232–233). TB2 used it and explained how she developed such an idea in her current teaching.

Interviewer: You assigned students to do projects this morning. Did you use the projects often?

TB2: This year in calculus I will. I will use probably one each quarter of four, during the course of the year.

Interviewer: Do you use projects for other courses?

TB2: No.

Interviewer: So why do you use, I mean why do you think it important for students to do projects in calculus?

TB2: Because, at the calculus level, I expect that they will be able to synthesize a great deal of their knowledge, to put it all together. And so every once in a while I want them to do that for me. In this first project is their synthesizing their knowledge so far on integrals, derivatives and limits. And then the next project I'm going to give will actually be a paper. They'll have to write any kind of paper, some of them choose to write poetry or short stories, on the meaning of the derivative. Relative to the second derivative and back and forth. So, that I expect from them and I expect them to be able to be responsible enough to get it done in a timeline.

Interviewer: So, when did you get this idea to use projects?

TB2: In my calculus classes in the past, I've assigned papers and some projects. So, I knew that I wanted to do it again this year. But, also with the reform calculus, from the information that I gathered from the conferences I had, I knew that this kind of synthesis of ideas was gonna be integrated to the new program.

---

[6]The actual name is not disclosed here for purposes of anonymity.

Table 7.7 Teaching strategies and the knowledge sources (TB2)

Teaching strategies	Sources identified
1. Cooperative learning	B: preservice training; G: own teaching experience and reflection
2. Assign students to do projects	D: organized professional activities; G: own teaching experience and reflection
3. Ask students to write journals in mathematics	D: organized professional activities; E: exchanges with colleagues
4. Use the overhead and clean and then rewrite over the transparency	G: own teaching experience and reflection

Table 7.7 presents the teaching strategies identified with TB2 and the knowledge sources she described.

Interview with TB3

As mentioned in Chapter 6, in one geometry class I observed, TB3 asked students to help her hold a ruler and make a right angle to demonstrate the concept of a right angle. The interview revealed this activity also integrated part of her general teaching strategy knowledge: involving students.

> Interviewer: How did you decide to ask students to hold a ruler and make a right angle [in the class]?
>
> TB3: When I prepare a lesson and if I think today I will have time, then I let a student participate. The kids like that. So, my goal is constantly to involve students as many different ways as I can.
>
> Interviewer: So when did you begin to notice that or have such a belief or philosophy?
>
> TB3: Well through the years, educationally, they keep telling us do this, do this, do this.
>
> Interviewer: Who?
>
> TB3: My department chairpersons. Every time the NCTM adopted a new policy, my boss at that time would say, try to do that. Never did they say, stop doing something. So consequently, I'm doing everything.... And after a while, you like some of the new things. And some of the ones you don't like, you don't use anymore. Next year you say, no, it didn't work; it was a waste of time. And you build up. It's just like a professional dancer. You build up a rapport and a repertoire

of what works. And you know what works. You know what doesn't. And you try for the best and the most.

TB3 is the only one I observed that directly required students to read the textbooks in both classes. It is interesting to note, during the interview, she admitted that she did not exactly know why she formed such a teaching strategy. (However, according to her further explanation, I coded the main source as "her own teaching experience and reflection".)

> Interviewer: Now I still want to talk about the classes I observed. When assigning the homework, in the first geometry preparatory class, you asked the students to study.
>
> TB3: I always do.
>
> Interviewer: Yes. I mean, study means what? Here it means reading or writing or doing questions? I'm not sure.
>
> TB3: Actually, study just means read. When I have "study", that means I want them to open up to chapter one and look at the material.
>
> Interviewer: OK. I've got it. In the second class I observed, you also reminded students to remember to read each section as they progress. The question to me here is why you focus so much on reading?
>
> TB3: Because if I give — inadvertently, I forget to mention a vocabulary word. And then for homework, it's in there. And they come back the next day and say, but you didn't show us that word. I'll say, I don't have to show you everything. I told you, you have to read the book.... [In addition] I stress reading all the time because I am not the one person that they should only believe. They should be able to read the book. They should be able to talk among themselves.
>
> Interviewer: So, the major source for you to develop such an idea is your teaching experience?
>
> TB3: Probably.
>
> Interviewer: Were you influenced by any other sources?
>
> TB3: I don't know how I came about it. Now there are very few people that I would look to for guidance because what I would look for now is more variety.

Table 7.8 presents the teaching strategies identified with TB3 and the knowledge sources she described.

Interview with TC1

The interview with TC1 about his PIK started with cooperative learning, and he clearly pointed out he learned it from his preservice training.

Table 7.8 Teaching strategies and the knowledge sources (TB3)

Teaching strategies	Sources identified
1. Involve students	E: exchanges with colleagues; G: own teaching experience and reflection
2. Elicit students' thinking and ask them why	C: inservice training; G: own teaching experience and reflection
3. Ask students to write journals in mathematics	D: organized professional activities; E: exchanges with colleagues
4. Ask students to read the textbooks	G: own teaching experience and reflection

Interviewer: Now I would like to ask you some questions about the general teaching strategy you use in your classroom. In algebra class [I observed], you used cooperative learning. You asked students to sit together working with a partner. How did you develop your knowledge of cooperative learning? I mean, that's a relatively new teaching strategy.

TC1; Yes. ... In my college education courses, they stressed cooperative learning.

Interviewer: When?

TC1: I did my education courses just three years ago. They stressed a lot of cooperative learning.

Interviewer: Which college?

TC1: [The college name]. So that concept came from there.

"Volunteering" has been recognized as a motivational strategy by classroom researchers (e.g., see Borich, 1992). TC1 explained that he learned that strategy when he did his previous training, but not from his educational courses; instead, it is from his classroom observation required for the preservice training program.

Interviewer: In the algebra class, you asked students to be volunteers to answer Questions 29 to 32. And later you asked a volunteer to give an example about equal fractions too.[7] In the precalculus class, you asked the students voluntarily again to give you an example of a quadratic function.[8]

---

[7] The example the student gave was $\frac{2}{4} = \frac{1}{2}$, which the teacher used to explain why $\frac{ak}{bk} = \frac{a}{b}$.

[8] A student gave the function: $f(x) = 4x^2 + 2x + 0$, then the teacher used the function to show how to find the vertex and then choose an appropriate window in a graphing calculator to graph the function.

Findings of the Chicago Study (III): Pedagogical Instructional Knowledge

Table 7.9 Teaching strategies and the knowledge sources (TC1)

Teaching strategies	Sources identified
1. Cooperative learning	B: preservice training
2. Ask students to read the textbooks	B: preservice training; G: own teaching experience and reflection
3. Have students voluntarily answer questions, give examples, etc.	B: preservice training
4. Generally involve students	E: exchanges with colleagues

TC1: That's where they're [the students] trying to trick you.

Interviewer: I don't know. To me, [my question is] how did you develop such an idea?

TC1: When I was doing my education courses three years ago in [the college name], you had to do 100 hours of observation. Now, I observed a very good teacher at Addison Trail. And one of the good things he did — he was teaching Algebra I. And one of the things he did with those kids a lot was problems on the board. Now he would pick volunteers. He'd say, you, you, you up on the board. So I watched him and I thought, well, this is a good idea to get the kids up there.

Table 7.9 shows the teaching strategies identified with TC1 and the knowledge sources he explained.

Interview with TC2

As noted in Chapter 5, TC2 had worked as an editor involving textbook development in a publishing company. The interview with TC2 proved that experience played a role, directly or indirectly, in her developing some relatively new teaching strategies. First let us see how she developed her knowledge of cooperative learning.

Interviewer: I noticed you used cooperative learning a lot. I wonder when was the first time for you to know the strategy.

TC2: Well, I think the first time was in [a] somewhat formal way. I used partners before in my first 11 years of teaching, but only like to review or as a study buddy kind of things. As far as the group of four in a little bit formal fashion, I think it was I learned the process when I was working in order to teach teachers, the method came in when I was not teaching, but to teach teachers the method, so I had to learn it. I had to read all about cooperative learning and so I learned about cooperative learning.

The second thing was I wanted to devise a method in my class that would make things very efficient. So, I would not have to answer many questions, and I thought well, in order for me to do that, they need to talk to somebody to get the questions answered and that was a way to do it.

The third reason I think was because I had a workshop on cooperative learning[9] that was held here and I volunteered to go to that workshop.

Interviewer: When?

TC2: Last January, maybe. I was a teacher here, using a little cooperative learning, but that really clinched it, when I went to that workshop.

Interviewer: So all those sources helped you develop your knowledge of cooperative learning.

TC2: Yes. But my first was because I had to teach it and I had to read about it.

In addition to her business experience as a textbook editor, TC2 also gave credit to professional organizations when talking about how she developed her knowledge of using "real application problems" in classroom teaching.

Interviewer: Well, how did you get to know giving real situation problems (e.g., O'Hare Parking Garage problem[10]) will help students learn better?

TC2: One reason that comes to mind is I worked with [the new textbook project][11] and that is a very, very strong aspect of [that project]. A second reason is that my professional organizations are saying this. They are saying "introduce topics with real world", "do this; develop through this".

Interviewer: And you read it from the publications [of the organizations]?

TC2: Yeah. I read it. I've heard it in talks. I've seen it in materials, not only in journal articles, but materials that have been written by these organization as sample problems. I heard it not just from form of seminars, but with colleagues who are in these professional organizations who are maybe research mathematicians in math education and they say it's better.

Interviewer: All of them helped you?

TC2: Yeah. In one way or another.

---

[9]TC3 in the same school also mentioned that workshop. Because here the workshop is special on cooperative learning, and has a purpose of training teachers (see "Interview with TC3" below), it was coded below as "inservice training", instead of "organized professional activities".

[10]See a description in Chapter 6.

[11]The actual name of the project is not disclosed here for purposes of anonymity.

Table 7.10 Teaching strategies and the knowledge sources (TC2)

Teaching strategies	Sources identified
1. Cooperative learning	C: Inservice training; D: organized professional activities; F: reading professional journals and books; G: own teaching experience and reflection
2. Discovery learning	E: exchanges with colleagues; F: reading professional journals and books; G: own teaching experience and reflection
3. Use real applications	D: organized professional activities; E: exchanges with colleagues; F: reading professional journals and books Other: business experience
4. Ask students to do oral presentation	A: experience as student; Other: business experience
5. Ask students to do projects	A: experience as student; D: organized professional activities; E: reading professional journals and books; Other: business experience

Table 7.10 shows the teaching strategies identified with TC2 and the knowledge sources she pointed out.

Interview with TC3

As said before, TC3 had been teaching for 20 years. Nonetheless, according to her own recall, she only started to use cooperative learning as one of her teaching strategy several years ago. She attended organized professional activities, received inservice training, and experimented with it herself.

> Interviewer: Regarding general teaching strategies, I noticed that in both classes you used a lot of cooperative learning. My question is when did you come to know about cooperative learning?
>
> TC3: I would say maybe over the last four or five years. I started experimenting with it, trying it in the classroom, and seeing that they are successful and end up learning more.
>
> Interviewer: Well, when was the first time for you to know?
>
> TC3: I'd say maybe four or five years ago. That's when I started doing it.

Interviewer: Where? And how?

TC3: Where? Oh, here at school, trying it on my own. And then some of ICTM workshops would have a speaker talk about it. And I'd go to listen to the speaker and get some ideas. Last year, our department chair brought in [an expert on it][12] to actually train us in cooperative learning.

A general teaching strategy TC3 used was letting students be in charge of their learning. This was particularly identifiable in the geometry class I observed, in which the teacher acted only as a guide. When asked why she had such a teaching strategy, TC3 clearly responded that it was a result of her teaching experience.

Interviewer: The geometry class I observed was devoted to questions; and questions were used to involve students. How did you design the strategy?

TC3: Students do better when they're in charge of learning. When they're presenting to each other and actively involved, they learn much more by problems on the board than if I just put them up myself on the board.

Interviewer: How did you get such kind of knowledge?

TC3: Through years of teaching. So it's by being an experienced teacher that you can see which methods work and which methods don't work.

TC3 also explained that she learned how to become a better teacher even from being a parent of her own children, when she was asked what was the general principle guiding her teaching practices. The general principle she mentioned is "to present the material in several ways".

Interviewer: Overall, what was the general principle guiding you?

TC3: Students learn in different ways and it's an obligation as a teacher to present the material in different ways, so that each student has a chance to learn. Some students might learn visually. Some might learn by hearing things. Another might learn by talking about it. One reason I have them put it on the board is they need to get up. They need motion. They can't sit there for 55 minutes. So some students learn better when there's an activity involved. If you present a wide variety of techniques, you've got a better chance that one of the techniques will work for the student.

Interviewer: OK. Now, how did you develop such a kind of philosophy?

TC3: By seeing how different students over the past years performed.

---

[12]The actual name is not disclosed here for purposes of anonymity.

Findings of the Chicago Study (III): Pedagogical Instructional Knowledge

Table 7.11 Teaching strategies and the knowledge sources (TC3)

Teaching strategies	Sources identified
1. Let students be in charge of learning	G: own teaching experience and reflection
2. Cooperative learning	C: inservice training; D: organized professional activities; G: own teaching experience and reflection
3. Let students help each other	C: inservice training; G: own teaching experience and reflection
4. Ask students to do writing	D: organized professional activities; F: reading professional journals and books; G: own teaching experience and reflection
5. Present materials in different ways	G: own teaching experience and reflection; Other: parenting own children

Interviewer: This is your teaching experience?

TC3: Right. And from my own children. I have three teenagers; they were very different students. So as they grew up and went to school, I saw the type of learners they were. That helped me be a better teacher as I could see them struggle and I could know what worked and what didn't work.

The general teaching strategies and techniques identified with TC3 and the knowledge sources described by her are presented in Table 7.11.

Table 7.12 provides a summary of the data from the interviews with the nine teachers from the perspective of how different sources were described by teachers for them to develop their knowledge of different teaching strategies and techniques.

From the table as well as the above discussion, it is clear that there are various sources by which teachers learn their knowledge of different teaching strategies and techniques. More specifically, on the one hand, the same teacher can learn his/her different teaching strategies and techniques from different sources, for example, the table below tells us that TC2 learned her different PIK from seven different sources.

On the other hand, for some common strategies, different teachers can also obtain their knowledge of them in different ways. In fact, the interview revealed that some teachers learned, for example, cooperative learning, from their "preservice training" (e.g., TA1, TB2, and TC1), while others

Table 7.12 Sources of teachers' PIK (from interview data)

Sources[a]	A	B	C	D	E	F	G	Other sources
TA1	yes (2)	yes (5)			yes (1)	yes (1)	yes (3)	
TB1	yes (1)				yes (2)		yes (3)	life experience (1)
TC1		yes (3)			yes (1)		yes (1)	
TA2		yes (1)			yes (3)	yes (1)	yes (2)	
TB2		yes (1)		yes (2)	yes (1)		yes (3)	
TC2	yes (2)		yes (1)	yes (2)	yes (3)	yes (3)	yes (2)	business experience (3)
TA3	yes (2)						yes (4)	
TB3			yes (1)	yes (1)	yes (2)		yes (3)	
TC3			yes (2)	yes (2)		yes (1)	yes (5)	parenting (1)
Total	4 (7)	4 (10)	3 (4)	4 (7)	7 (13)	4 (6)	9 (26)	3 (5)

Note: The figures in parentheses are the numbers of teaching strategies and techniques that teachers reported that they learned (regardless of how much) from the corresponding sources.

[a] A = Experience as student, B = Preservice training, C = Inservice training, D = Organized professional activities, E = Informal exchanges with colleagues, F = Reading professional journals and books, G = Own teaching experience and reflection.

learned from "inservice training" (e.g., TC2 and TC3), and still others learned from their "informal exchanges with colleagues" (e.g., TA2) and "experience as student" (e.g., TA3).

Moreover, besides all the seven sources investigated in the questionnaire, teachers also gained their knowledge from their other experiences such as daily life experience, previous business experience, and parenting their own children. Nonetheless, although there might be even more other sources, it is largely clear that the seven sources identified in the framework of this study did constitute a rather complete list of major sources. (Note: TC2's business experience is rather unique, as mentioned in Chapter 5.)

In contrast, it is largely unclear from the table how those different sources are comparatively important. Although it seems reasonable to say that teachers' own teaching experience and reflection and daily exchanges with their colleagues are the most important ones, which is consistent with what we found from the questionnaire data, it seems unwise to decide the relative importance of other sources just from the table, not just because the data in the table seem so close, but also because the interview sample consists of only nine teachers. Unlike the questionnaire, it is not the purpose of the interview to tackle the issue of relative importance.

Summary of the Findings

This chapter reports the findings about how teachers develop their PIK. In the study, the question is investigated through a survey questionnaire, which is intended to obtain general ideas about the variety and the relative importance of different sources by which teachers develop their PIK; classroom observation, which is to identify what teaching strategies and techniques teachers actually use in teaching; and interview, which is based on the classroom observation and performed in order to get specific information and contexts about how teachers develop the identified knowledge.

There are three main findings from the analysis of the data collected from both the questionnaire and interviews.

1. There are various sources by which teachers develop their knowledge of general teaching strategies and techniques, namely, their PIK. The same teacher can learn his/her different teaching strategies and techniques from different sources, and the same strategy can be acquired by different teachers through different ways.
2. Comparatively, teachers' "own teaching experience and reflection", "informal exchanges with colleagues", and "inservice training" are the most important sources for teachers to enhance their PIK; their attending "organized professional activities" is a secondarily important source; while their "preservice training", "experience as student", and "reading professional journals and books" are the least important sources.

Statistically, the importance of these three groups of sources to teachers' development of their PIK are significantly different.
3. The length of teachers' teaching experience does not play a significant role in teachers' viewing of the importance of different sources to the development of their PIK. In other words, each of the different sources carries basically the same importance to all teachers in developing their PIK, regardless of when they entered into their teaching career.

Chapter 8

Findings of the Chicago Study (IV): Some Other Issues

The previous three chapters presented the core findings of this study about how teachers develop their pedagogical curricular knowledge (PCrK), pedagogical content knowledge (PCnK), and pedagogical instructional knowledge (PIK), respectively, from different sources.

This chapter supplements the three previous ones by looking at some other issues, which seem not appropriate to be included in the previous chapters, but are related and helpful for us to understand the issues addressed in those chapters from a different angle. Different from Chapters 5–7, this chapter does not separate the general pedagogical knowledge into three different components, since the focus of this chapter is on specific sources. Nonetheless, it should be noted that when I discuss pedagogical knowledge in this chapter, it does not always include or apply to all its components. Distinctions can be made, if necessarily, according to the categorization of pedagogical knowledge aforementioned.

The chapter consists of three sections. The first section mainly addresses the issues about how teachers used or acted on specific sources by looking at different sources *individually*. For example, how often did teachers attend organized professional activities, and how useful did they find this particular source? The sources being examined are experience as student, preservice training, inservice training, organized professional activities, informal exchanges with colleagues, reading professional books and journals, among others.

The second section of this chapter, under the subtitle of "How teachers improve their pedagogical knowledge", is reserved for some questions included in the interviews but not reported in previous texts of this study. The first question asked if there had been major changes in their teaching strategies or styles recently, and if so, what the knowledge sources for such changes were. The second asked what the major sources of those teachers' pedagogical knowledge overall in their teaching careers were. The third asked what they thought were the main difficulties in improving teachers' pedagogical knowledge and how to pursue the improvement. The final section of the chapter summarizes the main points of the wider issues found in the study.

As in Chapters 5–7, the findings reported in this chapter are based on data obtained from the questionnaire survey and/or interviews.

How Teachers Use Different Sources

Experience as students

As we can see from the findings discussed in Chapters 5–7, teachers' experience as students is, comparatively, one of the least important sources for teachers to develop their pedagogical knowledge, except for their knowledge of textbooks, and PCnK, for which it is a secondarily important source. Given the crucial influence of one's knowledge on his/her behavior, this finding implies that what people often claimed that "teachers teach the way they were taught" is likely not true.

Question 5 of the questionnaire of this study (see Appendix 1A) provides another way to look at the issue. The question asks which teaching strategies they were exposed to as students and what they used as teachers. The main result is summarized in Table 8.1.

The table shows that, while there is no or only a little difference between the percentage of teachers who used reviewing, lecturing, and students' reading, respectively, in their teaching and that of teachers who were exposed to the corresponding strategies as students, there exist great differences for using computers and calculators. The differences for using small group work (a sign of cooperative learning), concrete materials, hands-on activities, and classroom discussion are also considerable. The reasons for the differences seem obvious, as computers, calculators, and, to

Table 8.1 Teaching strategies used as teachers and encountered as students, reported by teachers

Teaching strategies*	% used as teachers	% encountered as students	% difference
a. review of previous lessons (T = 68, S = 59)[†]	100	94.9	5.1
b. lecture on new topics (T = 68, S = 68)	100	100	0
c. classroom discussion (T = 69, S = 65)	97.1	70.8	26.3
d. small group work (T = 69, S = 67)	98.6	40.4	58.2
e. reading assignment from the texts (T = 68, S = 66)	75	75.8	−0.8
f. use of calculator (T = 69, S = 68)	100	35.3	64.7
g. use of computer (T = 69, S = 60)	94.2	13	81.2
h. use of concrete materials (T = 69, S = 63)	95.7	46	49.7
i. hands-on activities (T = 69, S = 66)	95.7	40.9	54.8

*The teaching strategies listed below are by no means exhaustive, or exclusive with each other. Instead, they are commonly used and discussed teaching strategies.
[†]T and S are the numbers of teachers who answered if the teaching strategy was used by themselves (as teachers), and encountered as students, respectively. The numbers are not necessarily 69, as some of the teachers chose "don't remember" or did not respond to that source.

a lesser degree, concrete materials are nowadays increasingly available and encouraged to be used in classrooms, and cooperative learning, hands-on activities, and classroom discussion are also relatively new or receive more emphasis mainly for the purpose of involving students in the learning process.

There is little doubt that many teachers were not exposed to a number of teaching strategies during their school education, so their experiences as students could not be a main source for them to get their knowledge of those strategies.[1] Therefore, the results displayed in Table 8.1 are consistent

---

[1] Note: Question 5 of the questionnaire only considers experience as students in mathematics classes. As shown in Chapter 7, it is possible that a teacher acquired their pedagogical knowledge from their school experience in other subjects. However, teachers reported such transfer only rarely.

with what we found previously about the importance of that source to the development of teachers' pedagogical knowledge.

It is interesting to note that, in the interviews, among all the 12 teachers (including the three math chairs) who were asked the following question:

> Some people say that teachers teach the way they were taught. Do you agree with the statement or not? [Note: college experiences were explicitly included]

Eight (66.7%) of them answered "no" to the question (one teacher even stated that "I teach the way I was not taught"), two (16.7%) said "initially yes, but later no", and only two of the remaining (16.7%) with relatively shorter teaching experience claimed "generally yes". Let us take a look at how three of the teachers responded:

> TA2: In general, yes. But in specific, I guess the thing I have a problem with is that if I look back at my teachers in high school when I was being formed, I had a different teacher every year that didn't necessarily use the same style.
>
> TC1: I think we work it the way we were taught first and then we try to improve on it. I never used a calculator [when I was at school]. I wasn't taught with a calculator, so that's new. That wasn't the way I was taught. But when I get down and say, let's complete this square, that was the way I was taught. So I'm still bringing back that idea, but I'm showing them why it's important now.
>
> MC2:[2] I think teachers teach the way they were trained to teach. Now for my generation we were probably trained to teach in much the same way we were taught and that was by lecture, but as you keep developing that training how you teach changes. What I see teachers coming out with now when they are graduating from college is very different, they are trained in constructive approach, discovery learning and cooperative learning, so forth. So, I think it's depending on how they were trained when they were in college.
>
> I think when you start out your first days on the job it's going to be what you learned in preservice training and what you learned during student teaching. If you had a cooperating teacher in student teaching or professor when you did your inservice work who didn't expose you to a variety activities, you're not going to incorporate a variety of activities in your classroom. If then, given your start you are exposed to other ideas either through working with other teachers in the building or your chair or workshops then you may start incorporating those in your teaching.

---

[2] MC2 refers to the math chair in the second school, i.e., School 1B.

When further asked what the differences between the way they were taught and the way they taught are, most teachers listed that they used more technology, more cooperative learning, and more methods of engaging students, consistent with the result shown in Table 8.1.

MC2's statement that "teachers teach the way they were trained to teach" is particularly enlightening here, if we change "were trained" to "know", since teacher training, especially preservice training actually is not an important source for teachers to get their pedagogical knowledge as discussed in previous chapters. Moreover, teachers' experiences as students are overall the least important source of their knowledge of how to teach, so how can it be true that teachers are only bound to the knowledge from a minor source?

In short, based on the results pointed out above as well as the fact revealed in the previous chapters, the notion that teachers teach the way they were taught is, at most, only true for a very small proportion of teachers in this population, and in general, the statement is not supported by the findings of this study from the teachers in these schools.

Preservice training

According to teachers' responses to Question 6 of the questionnaire, 66 (95.7%) teachers received preservice training. Among them, 63 (95.6%) had training for teaching mathematics, both rates being very high as expected since they are in the best schools.

Figure 8.1 presents the percentages of teachers who were taught various pedagogical knowledge (skills), based on teachers' responses to Question 7 of the questionnaire, with the average response rate being 93.3% (s.d. = 8.9%), which excludes those answering "don't remember".

It is surprising that, according to the figures, a considerable portion of teachers did not receive necessary training on some basic pedagogical knowledge (skills) in their preservice training programs; it turns out that clearly those teachers have to resort to other sources to develop their pedagogical knowledge which they are not taught in preservice training.

In addition, the figure shows that the lowest percentages (from 12.1% to 22.7%) of teachers, who more likely are younger teachers, had training

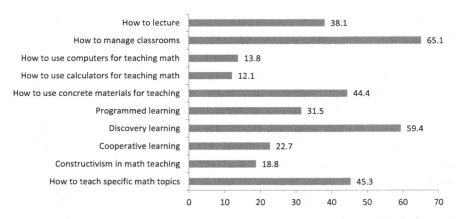

Figure 8.1 Percentages of teachers who were taught different pedagogical knowledge in their preservice training

in using technology, cooperative learning, and constructivism in mathematics learning. Considering that nearly 50% of the teachers in the sample of this study have 15 or less years' teaching experience, and about 25% have 5 years or less (see Table 1E.5 in Appendix E), I think those percentages to some degree imply that preservice training programs need to be improved so that more and newer developed pedagogical knowledge and skills can be integrated into the training courses and introduced to prospective teachers.

Conventionally categorizing preservice training courses into general educational courses, mathematical methods courses, and student teaching practice, Question 8 of the questionnaire asks teachers how useful they feel those courses were in enhancing their pedagogical knowledge.

Table 8.2 shows the results of teachers' evaluation of those three kinds of courses. It is evident from this table that all the teachers except one considered their student teaching experience "very useful" (74.6%) or "useful" (23.8%). Less than 60% believed the mathematics methods courses they took in preservice training were "very useful" (15.3%) or "useful" (42.4%). Less than one-quarter considered the general educational courses they took "very useful" (3.1%) and "useful" (20%). That is, nearly 77% of the teachers felt general educational courses not useful (40% being not very useful and 36.9% not useful). Combining those percentages with the average scores, we can see that student teaching is definitely much more

Table 8.2 Teachers' evaluation of the usefulness of preservice training courses in enhancing their pedagogical knowledge

	Very useful	Useful	Not very useful	Not useful	Average*
General educational courses (n = 65)	2 (3.1%)	13 (20%)	26 (40%)	24 (36.9%)	1.89
Mathematics methods courses (n = 59)	9 (15.3%)	25 (42.4%)	19 (32.2%)	6 (10.2%)	2.63
Student teaching practice (n = 63)	47 (74.6%)	15 (23.8%)	1 (1.6%)	0 (0%)	3.73

*The average score is obtained by using the ordinal scale, 4 = very useful, 3 = useful, 2 = not very useful, and 1 = not useful.

important than mathematics methods courses, which in turn are considerably more important than general educational courses, in terms of equipping preservice teachers with pedagogical knowledge.[3]

Five of the nine teachers who were interviewed explained the importance of student teaching. When talking about her preservice training during the interview, TC3 described,

> The preservice training in college I feel was very poor. Wonderful was student teaching. But our methods class really didn't do enough to prepare us. And I didn't have any observing before I student taught. But I think to prepare teachers, student teaching is the most valuable experience you can have. (Interviewer: So you learned what from your student teaching?) You learned classroom management. You learned what students were like. You learned some of the things that would work with students and what didn't work with students. I think I would like it — if I had a wish it would be that instead of waiting until second semester of senior year for student teaching, they should have that you have a semester where you observe or teacher aide, you know, hook up with a high-school classroom teacher and get some more knowledge, some more observation hours.

TA3 said,

> I didn't learn much [from my preservice training]. I mean, student teaching was good. But the classes? I didn't learn much from my classes in college. I learned

---

[3]Applying the chi-square test to the data between the general educational courses and mathematics methods courses shows, as expected, a significant difference, with $\chi^2 = 19.8870$, df = 3, and p = 0.00017. To the data between mathematics methods courses and student teaching, the p-value is less than 10^{-10}.

math, but I didn't learn much about education. Student teaching was very good. Student teaching is very beneficial. People should do more student teaching and less classwork.

Back to the questionnaire, Question 9 asks teachers to give an overall evaluation of the usefulness of their preservice training to developing their knowledge of how to teach mathematics. The result from 65 teachers who answered the question shows that 10.8% of them considered their preservice training "very useful", 43.1% considered it "useful", while 36.9% considered it "not very useful", and 9.2% rated it "not useful".

Overall, only a little bit more than 50% of teachers hold positive evaluation on their preservice training. By using the ordinal scale, 4 = very useful, 3 = useful, 2 = not very useful, and 1 = not useful, the average evaluation score was 2.52, a very low level compared to other sources.

Inservice training

Questions 10–12 of the questionnaire relate to inservice training. According to teachers' responses, 71% (49) took college/university courses for credit but not for degree (non-degree training program) and 66.7% (46) attended degree training programs after the first time they became teachers.

Table 8.3 classifies courses taken by the 49 teachers into different fields, while Table 8.4 displays the information about the majors the 46 teachers enrolled in in degree programs.

For those having received inservice training, 96% attended non-degree inservice training by their own choice, and the figure for degree programs is 98%. In contrast, only a small number of teachers also explained that they attended either non-degree or degree inservice training programs partially or solely because of the district (10% for non-degree programs, 11% for degree programs) or state requirement (2% for each kind of programs), or for other reasons (14% for non-degree, 20% for degree) — dominantly salary enhancement. Those results suggest that there were generally no strong administrative stimulants or requirement for teachers to seek further university-based professional training after they finished their preservice training.

Table 8.5 displays teachers' evaluation of non-degree and degree inservice training programs based on the data obtained from Questions 10 and 11.

Table 8.3 Courses taken by teachers in non-degree professional training

	Courses taken*				
	Technology	Mathematics	General education	Teaching and mathematics teaching	Other
No. of teachers (n = 49†)	20	30	7	12	4
% of all teachers	40.8%	61.2%	14.3%	24.5%	8.2%

*Technology — including calculator, computer, and technology use in classroom. Mathematics — including applied calculus, discrete mathematics, mathematics history, etc. General education — including educational administration, classroom management, discipline, multicultural education. Teaching and mathematics teaching — including mathematics education, cooperative learning. Other — including science, athletic training, writing, engineering.
†Some teachers took more than one kind of course.

Table 8.4 Distribution of the majors teachers enrolled in inservice degree programs

	Majors* enrolled in			
	Mathematics	Mathematics education	General education	Other
No. of teachers (n = 46†)	12	21	8	6
% of teachers	26.1%	45.7%	17.4%	13%

*Mathematics — including both mathematics and statistics. Mathematics education — including MST, MAT, curriculum and instruction, and educational psychology in mathematics. General education — including educational administration, secondary education, educational and social policy, etc. Other — including computer, administration, recreation administration, physical education, etc.
†In all the 46 teachers who have enrolled in one or more degree programs after being teachers, one was for a bachelor's degree, two were for doctoral degrees, and all the others were for master's degrees. One was enrolled in two master's programs.

The table illustrates that between 75% and 80% of teachers considered the inservice training they received, whether for an advanced degree or not, useful or very useful. Moreover, comparing the average scores, which are very close to each other, to that for preservice training aforesaid, it is clear that inservice training is more important than preservice training in enhancing teachers' pedagogical knowledge, which is consistent with the results reported in Chapters 5–7.

Table 8.5 Teachers' evaluation of the usefulness of inservice training programs in enhancing their pedagogical knowledge

	Very useful	Useful	Not very useful	Not useful	Average*
Non-degree programs	15	23	7	4	3
(n = 49)	(30.6%)	(46.9%)	(14.3%)	(8.2%)	
Degree programs	13	21	4	7	2.9
(n = 45[†])	(28.9%)	(46.7%)	(8.9%)	(15.6%)	

*The average score is obtained by using the ordinal scale, 4 = very useful, 3 = useful, 2 = not very useful, and 1 = not useful.
[†]Among the 46 teachers who had enrolled in degree inservice training programs, one had just begun the program and gave no evaluation.

Question 12 of the questionnaire looks further into what kinds of other professional training teachers received, and how useful they are. Teachers' answers show that 94.2% (65) of the teachers received inservice training focusing on using computers or calculators for teaching mathematics (PCrK, technology), and 75.4% (52) received training on new teaching methods and strategies (PIK).

A remarkable contrast is that the figure for inservice training focusing on how to teach particular mathematics topics (PCnK) is 40.6% (28) and that on textbooks (PCrK, textbooks) and other teaching resources is only 21.7%, which in a sense explains why teachers regarded "inservice training" as one of the most important sources for them to develop their PCrK in technology and PIK, but not for PCnK and other parts of PCrK, as discussed in Chapters 5–7.

Table 8.6 depicts the teachers' evaluation of how useful those other types of inservice training are. From the table, we can see that no one chose "not useful" for those types of inservice training, and only a very few teachers chose "not very useful". Furthermore, based on the average scores, inservice training on using technology and on new teaching methods and strategies are more useful than those on how to teach particular mathematics topics and on textbooks and other teaching materials, which again is consistent with the results reported in the previous three chapters.

Almost all (91%) teachers reported that they attended those professional training sessions of their own choice, 23% reported partially or solely because of district requirement, 12% for other reasons such as

Table 8.6 Teachers' evaluation of the usefulness of other professional training in enhancing their pedagogical knowledge

Training focusing on	Very useful	Useful	Not very useful	Not useful	Average*
textbooks and other teaching resources (n = 15)	4 (26.7%)	10 (66.7%)	1 (6.7%)	0	3.20
using computer/calculator for teaching mathematics (n = 65)	40 (61.5%)	19 (29.2%)	6 (9.2%)	0	3.52
new teaching methods and strategies (n = 52)	26 (50%)	24 (46.2%)	2 (3.8%)	0	3.46
how to teach particular mathematics topics (n = 28)	11 (39.3%)	14 (50%)	3 (10.7%)	0	3.29

*The average score is obtained by using the ordinal scale, 4 = very useful, 3 = useful, 2 = not very useful, and 1 = not useful.

salary enhancement or department/school encouragement, and none for state requirement.

Organized professional activities

Questions 13 and 14 of the questionnaire are designed to obtain information on teachers' attending organized professional activities at various levels.

According to teachers' responses to Question 13, in the last five years they on average attended seven (s.d. = 3.7) professional activities at the local and state level, and two (s.d. = 6.4) at the regional or national level. The difference between the local/state level and the regional/national level is very considerable yet understandable, since attending regional and national professional activities usually takes more time and costs more.

All the 67 teachers who attended those activities (the other two teachers did not attend activities at either level) included "own choice" as a reason for their attending, 9% of teachers also listed "required by district" as a reason, and still 9% gave other reasons, such as "encouraged by department", "attended as a speaker", and so on.

Table 8.7 Teachers' evaluation of the usefulness of organized professional activities at the local/state and regional/national levels

	Did not attend	Not useful	Not very useful	Useful	Very useful	Average*
Local/State (n = 69)	2	2 (3%)	10 (14.9%)	41 (61.2%)	14 (20.9%)	3.00
National/Regional (n = 69)	21	0 (0%)	6 (12.5%)	31 (64.6%)	11 (22.9%)	3.10

*The average score is obtained by using the ordinal scale, 4 = very useful, 3 = useful, 2 = not very useful, and 1 = not useful.

Table 8.7 presents the teachers' evaluation of the usefulness of organized professional activities at the local/state and regional/national levels in enhancing their knowledge of how to teach mathematics.

Initial examination of the table shows that, although the average scores seem very close, about 18% of the teachers were disappointed in the usefulness of the professional activities they attended at the local/state level, and the figure for the regional/national level is only 12.5%. Thus, a greater proportion of teachers felt professional activities organized at the regional/national level were "useful" and "very useful". Applying the chi-square test to the data in the table results in $\chi^2 = 8.485$, df = 3, and p = 0.0370, statistically significant at the 0.05 level. In other words, teachers felt professional activities at the regional/national level are significantly more useful in enhancing teachers' knowledge of teaching mathematics than those organized at the local/state level.

Teachers' responses to Question 14 reveal that 74% of the teachers attended professional activities at the school/department level once per month or once per semester, and 12% once per year during the school year of 1996–1997; the percentages for the district/county level are 29% and 38% respectively. Also, 13% chose "other" for professional activities at the school/department level, and 33% chose that for activities at the district/county level. As almost all of those choosing "other" did not provide further information (only a very few added, e.g., "more than once per month", etc.), it is unclear how often those teachers attended professional activities at the two levels, though it is clear that, as expected, teachers did

Table 8.8 Teachers' evaluation of the usefulness of organized professional activities at the school/department and district/county levels

	Not useful	Not very useful	Useful	Very useful	Average*
School/Department (n = 62)	2 (3.2%)	11 (17.7%)	40 (64.5%)	9 (14.5%)	2.90
District/County (n = 47)	6 (12.8%)	23 (48.9%)	16 (34%)	2 (4.3%)	2.30

*The average score is obtained by using the ordinal scale, 4 = very useful, 3 = useful, 2 = not very useful, and 1 = not useful.

attend more professional activities at the school/department level than at the district/county level.

A more important issue here is how useful those professional activities are in teachers' developing their pedagogical knowledge. Table 8.8 presents the results of teachers' answers.

As seen from the table, nearly 80% of the teachers considered professional activities they attended at the school/department level useful or very useful. In contrast, the percent for the district/county level is less than 40%. Employing the chi-square test to the data shown in the table leads to $\chi^2 = 14.1385$, df = 3, and p = 0.0027, which means, based on teachers' evaluation, professional activities at the school/department level are viewed as statistically significantly more useful than those at the district/county level in enhancing their knowledge of teaching mathematics.

Further applying the chi-square test to the data on the local/state level in Table 8.7 and on the school/department level in Table 8.8, we get $\chi^2 = 0.6282$, df = 3, and p = 0.8900, which suggests there is no significant difference between those two levels.

To conclude, considering professional activities organized at all the four levels in Questions 12 and 13 together, the study shows professional activities at the regional/national level are felt to be significantly more important in developing teachers' pedagogical knowledge than those at the local/state level and at the school/department level, which in turn are significantly more important than those at the district/county levels. As teachers on average only attended two professional activities, and 21 (30%) teachers did not attend any professional activities, at the regional/national level in the last

Table 8.9 Frequencies of teachers having non-organized professional activities

	Never	Rarely	Monthly	Weekly or biweekly	2 or 3 times a week	Almost daily
Classroom observation	39	25	2	1	1	1
Informal exchanges with colleagues	0	1	1	8	17	42
Reading professional journals and books	1	16	29	19	3	1

Note: n = 69.

five years, the result implies that teachers should have more opportunities to attend national and regional professional meetings, workshops, and the like; also the quality of district/county professional activities should be particularly improved as they have played the weakest role in developing teachers' teaching knowledge compared to professional activities organized at the other three levels.

Non-organized professional activities

Question 15 of the questionnaire contains three non-organized professional activities that I believed would play important roles in teachers' daily professional life — classroom observation, informal exchanges with colleagues, and reading professional journals and books. The results are shown in Table 8.9.

The table shows that the overwhelming majority of teachers (93%) never or rarely observed their colleagues' classrooms during the school year. Moreover, one responding "weekly or biweekly" is a math department chair, who explained in the interview it was his responsibility as a department chair to observe classes regularly. Although it is not clear what are the reasons for the other four to have such activities from "monthly" to "daily" as the questionnaire survey is anonymous, and some might have special responsibility to do so, there is no doubt that overall teachers are still isolated within their classrooms, as pointed out decades ago (e.g., see Lortie, 1975).

With respect to "informal exchanges with colleagues", the table suggests it happened to most teachers (61%) on a daily basis, and to another 36% of the teachers from biweekly to two or three times per week.

As to "reading professional journals or books", 25% of the teachers never or rarely did so during the school year, and the remaining 75% did that from monthly (42%) to almost daily (the math chair aforementioned). In the interviews with 11 teachers,[4] eight (73%) of them agreed that teachers were too busy to read professional journals and books.

Among the professional journals they did read, *Mathematics Teacher*, published by the NCTM, was the one mentioned most. However, teachers also read other journals, as TB3 explained clearly,

> About the magazines — I don't know if you've ever heard of the magazine called *Games*. It's a very unusual magazine. It has mental puzzles, word puzzles, and mathematical puzzles. I like to read those kinds of magazines. And that's where I get some useful ideas for teaching. I read everything. And when I find something interesting, I try to use it in my classroom.

Back to the questionnaire, Table 8.10 displays the distribution of the numbers of teachers who gave different evaluations of the usefulness of those non-organized professional activities.

According to figures shown in Table 8.10, we can see that most teachers believed that all of the professional activities are useful or very useful, with the percentages for classroom observation, informal exchanges with colleagues, and reading professional journals and books being 67%, 99%, and 60%, respectively.

Applying the chi-square test to the data on the first two kinds of activities yields a p-value less than 10^{-7}, therefore, "informal exchanges with colleagues" are statistically significantly more useful than "classroom observation", and obviously "reading professional journals and books" as well, which is consistent with the findings reported in Chapters 5–7.

---

[4] One math chair who did not teach classes in the semester was not asked the relevant questions.

Table 8.10 Teachers' evaluation of the usefulness of non-organized professional activities in enhancing their pedagogical knowledge

	Not useful	Not very useful	Useful	Very useful	Average
Classroom observation (n = 30)	7 (23.3%)	3 (10%)	15 (50%)	5 (16.7%)	2.60
Informal exchanges with colleagues (n = 69)	0 (0%)	1 (1.4%)	14 (20.3%)	54 (78.3%)	3.77
Reading professional journals and books (n = 68)	7 (10.3%)	20 (29.4%)	38 (55.9%)	3 (4.4%)	2.54

*The average score is obtained by using the ordinal scale, 4 = very useful, 3 = useful, 2 = not very useful, and 1 = not useful.

Table 8.11 How teachers use different sources to design the ways to represent new mathematics topics

	Never	Rarely	Sometimes	Most of the time	Always	Average*
Textbook/teachers' notes	1 (1.5%)	3 (4.5%)	23 (34.3%)	26 (38.8%)	14 (20.9%)	2.73
Professional books or journals	11 (15.4%)	23 (35.4%)	32 (49.2%)	1 (1.5%)	0 (0%)	1.34
Exchanges with colleagues	0 (0%)	0 (0%)	20 (29.9%)	27 (40.3%)	20 (29.9%)	3.00
Your own knowledge	0 (0%)	0 (0%)	3 (4.5%)	15 (22.4%)	49 (73.1%)	3.69

*The average score is obtained by using the ordinal scale, 4 = always, 3 = most of the time, 2 = sometimes, 1 = rarely, and 0 = never.

Knowledge sources used in designing the way to represent mathematics topics

Finally, Question 22 of the questionnaire asks teachers that when teaching lessons introducing new mathematics topics during the school year, how they used the following four sources — textbook/teachers' notes, professional journals and books, exchanges with colleagues, and their own knowledge (see Appendix 1A), to design the ways to represent the new mathematics topics.

Table 8.11 depicts teachers' responses to the question.

It can be seen that, from the table, teachers most frequently used their own knowledge (which might have developed from their preservice training, inservice training, preservice teaching experience, and other sources) and cooperated with their colleagues to design the way to represent new topics. They also resorted to textbooks and teachers' notes. Comparatively, the resource they utilized the least among the four sources is professional journals and books.

Applying log-linear regression to the data, whose main result is attached in Table 1F.5 in Appendix 1F, we see that, according to the parameter estimates, the order of the frequencies of the sources teachers used to design the way to represent new mathematics topics is, their own knowledge (D: -2.4319), exchanges with their colleagues (C: -0.5422), textbooks and teachers' notes (A: 0), and professional journals and books (B: 3.7316).

Moreover, the Pr > Chi-Square values in the table reveal that "your own knowledge" is significantly more frequently used than "textbook/teachers' notes", there is no significant difference between the frequency of use of "exchanges with colleagues" and that of "textbook/teachers' notes", while "professional journals and books" is significantly less frequently used than "textbook/teachers' notes", all at the 0.05 level.

The above result is consistent with the findings discussed in Chapter 6.

How Teachers Improve their Pedagogical Knowledge

In the interviews, teachers were also asked to respond to three questions that were not reported previously but are related to the theme of the study.

The first question asked 11 of the 12 teachers (including two of the three math chairs)[5] if there had been major changes of their teaching strategies or styles recently (compared to five or ten years ago), and if so, what were the knowledge sources for such changes?

The three teachers at their early stage of teaching career answered, respectively, "not really", "nothing major, a little bit looser in classroom management", and "only a little bit, more flexible in classroom structures". For the latter two answers, the teachers both explained the changes were because of their getting more teaching experiences.

---

[5]The third math chair, not asked the question, did not have teaching duty in the semester.

All the other eight teachers said "yes" to the question. Six teachers stated that "using more technology" was a major change in their classroom teaching, and six also mentioned "cooperative learning" — four of them used the strategy more often now than earlier, one used it more selectively, and one changed the way to organize cooperative learning (students now always sitting in groups). Besides that, three teachers employed more student-centered approaches; three had more various ways to teach mathematics; one had now everything pre-prepared and got all transparencies done in advance; and finally one teacher integrated more applications.

Although teachers may not have disclosed all the major changes they had made, the above list proves that teachers did make changes. When those eight teachers were asked to identify the knowledge sources for such changes, four of them listed "their own teaching experience and reflection", three listed "colleagues", three "inservice training", two "professional meetings", and one "reading professional journals and books". Excluding their previous "experience as student" and "preservice training" as it is unlikely that those two sources continue to contribute to the changes, we see again that the above result is largely consistent with the findings reported in the previous three chapters.

The second question is what have been the major sources for teachers to develop their pedagogical knowledge in their teaching careers. All of the 12 teachers were asked this question in the interview. Ten teachers described their own teaching experience and reflection as the most important or one of the most important sources. For example, TC3 said:

> What became my major source of being a good teacher was to constantly read, observe and learn from my experiences. . . . My own experience was definitely number one. If I had a bad experience, then I immediately tried to find a remedy for it, either in the textbook, asking somebody else why isn't this working, or trying on my own something different next time. But you constantly have to be flexible. You cannot think that I am the best, I am the only one who's the best, only I know how to do it right. Because then you do not allow yourself to grow and to adopt someone else's theories.

TC1 explained more briefly:

> As far as improving my knowledge, it's just trial and error. That's what improves your knowledge. (Interviewer: So that's your own teaching experience?) Yes. The teaching experience is probably the most important.

Eight teachers also claimed that having everyday exchanges with their colleagues was one major source for their knowledge of teaching. TC1 offered a relatively more detailed response about the issue:

> [A colleague name][6] will say, where are you at in that class? What did you do yesterday? So you're usually on a daily communication with the people teaching other sections of your class. As far as the other teachers, you can only talk to them especially if you run into something where you say: OK, I ran into this, you're teaching the class before, is that [something] you should expect students to know or should not expect them to know? If you have a student that's giving you a problem, you know, he's not performing well, and you go back to his previous math teacher and say: OK, what works with this kid that I'm not hitting? [those exchanges with your colleagues are] Really helpful. I think that needs to be stressed to any new teacher that if you get a problem, and make sure you ask somebody in your department. Because don't think that you hit it for the first time.

TA2 provided another vivid picture of how he maintained daily exchanges with his colleagues and why he thought it was important:

> They [the exchanges] are lots of times not long conversations, but it's two minutes here, two minutes there, another two minutes here. But if you add it all up, it's a sizable amount of talking time and then afterwards, thinking time. It deals with what to teach and how to teach it, and maybe when to teach it; [it is about] everything.

Besides, most teachers were able to recall and tell a recent example of such exchanges to explain what pedagogical knowledge they acquired from the exchanges. As Chapters 5–7 contained many of such concrete examples, I shall not repeat them.

Other sources described as major are inservice training (six teachers), organized professional activities (six teachers), reading professional journals (three teachers), preservice training (one teacher), and learning experience (one teacher). Interestingly, like TB3 (see Chapter 7), TA3 also described parenting her own children as a major source of her knowledge of how to teach more effectively.

These responses are consistent with the findings discussed in the previous chapters.

---

[6]The actual name is not disclosed here for purposes of anonymity.

The final question of the interviews with all the teachers and the math chairs asked what they thought the main difficulties in improving teachers' pedagogical knowledge were and how to pursue the improvement.

As expected, most (75%) interviewees thought the biggest difficulty was the fact that teachers have heavy teaching loads and other duties, and are too busy to spend enough time and energy on seeking professional development. Four teachers believed teachers' lacking motivation was a major difficulty to overcome. Two teachers considered lacking "knowledgeable speakers" and "role models" as the most difficult thing for teachers to develop their new teaching strategies and skills. Nonetheless, no financial difficulty was raised in the interviews, for they were from schools with ample financial resources.

The interviewees also provided various suggestions for how to improve teachers' pedagogical knowledge. Following is the list of those suggestions by the nine teachers:

A. Schools give teachers more free time to seek professional development (three teachers).
B. Schools offer mentoring programs for new teachers (three teachers).
C. Schools provide good leaders and motivate teachers to learn continuously (three teachers).
D. Preservice training programs offer more time for student teaching (two teachers).
E. Schools invite good speakers to present new pedagogical knowledge and skills (one teacher).
F. Teachers need to attend more outside professional meetings, at least once a year (one teacher).

The three math chairs made the following statements, respectively, from the perspective of being math chairs.

G. Let teachers know clearly where you stand; create collaborations within the department; and hold teachers accountable for their teaching.
H. Place most effective teachers in leadership positions, and support them to share ideas [with colleagues].

I. Find resources to support teachers to seek professional activities, and have all teachers in one office so teachers have more opportunities to exchange ideas.

From both the teachers and the math chairs, we can see that most of the suggestions imply that teachers need more frequent and more effective exchanges with their colleagues, either formally such as mentoring programs (B), or informally such as creating collaboration in the department (G, H) and having teachers in one office together (I). The importance of good teachers' leadership is also highly recognized (B, C, E, and H).

Summary of the Findings

This chapter supplements Chapters 5–7 by examining some issues not included in the three chapters. The issues discussed in this chapter are more general (not separating teachers' pedagogical knowledge), but related to and helpful in understanding the main findings reported in the previous three chapters.

The first section of this chapter addresses the questions about how teachers used or acted on specific sources by looking at different sources individually. The sources being examined are mainly experience as student, preservice training, inservice training, organized professional activities, informal exchanges with colleagues, and reading professional books and journals.

The second section of the chapter discusses three questions dealing with the knowledge sources for teachers' major changes (if any) in their teaching practices, the major sources of teachers' pedagogical knowledge in their teaching careers overall, and the main difficulties in improving teachers' pedagogical knowledge and the ways to pursue the improvement.

Provided below is a summary of the main points, among others, established in this chapter according to the data from the questionnaire and interviews. The average evaluation score included in the summary below for each source is obtained by using the same ordinal scale, $4 =$ very useful, $3 =$ useful, $2 =$ not very useful, and $1 =$ not useful.

1. Experiences as students play a minor role in teachers building up their pedagogical knowledge. Overall, these teachers employed more

teaching strategies, especially using technology and some relatively new teaching strategies such as cooperative learning, than what they had been exposed to as school students.
2. A considerable portion of teachers did not receive necessary training on some basic pedagogical knowledge and skills in their preservice training programs. Teachers hold a very low evaluation (average score: 2.52) of the usefulness of their preservice training in enhancing their pedagogical knowledge. Student teaching is believed to be considerably more useful (average score: 3.73), and within the preservice training system, significantly more important than mathematics methods courses (average score: 2.63), which in turn are significantly more important than general educational courses (average score: 1.89) in enhancing preservice teachers' pedagogical knowledge.
3. More than 75% of the teachers believe that university-based degree inservice training programs and non-degree inservice training programs are useful (average scores: 3.0 and 2.9, respectively). Teachers also evaluate other kinds of inservice training favorably, with the average scores being 3.52 for inservice training they received on using technology, 3.46 on new teaching methods and strategies, 3.29 on how to teach particular mathematics topics, and finally 3.20 on textbooks and other teaching materials.
4. More than 90% of the teachers attended various professional activities of their own choice. Professional activities organized at the regional/national level (average score: 3.10) were reported by teachers as significantly more important in developing their pedagogical knowledge than professional activities organized at the local/state level (average score: 3.00) or at the school/department level (average score: 2.90), which in turn are significantly more important than activities at the district/county levels (average score: 2.30).
5. Almost all the teachers maintain informal exchanges with colleagues regularly, while 75% of the teachers read professional journals and books at least once a month, and very few teachers observe other teachers' classrooms. Teachers' daily exchanges with their colleagues are very useful in enhancing their pedagogical knowledge (average score: 3.77), and statistically significantly more important than "classroom

observation" (average score; 2.60) and "reading professional journals and books" (average score: 2.54).
6. Most teachers interviewed in this study have made major changes in the ways they teach in recent years. The main knowledge sources for teachers to make changes are their own teaching experience and reflection, daily exchanges with colleagues, inservice training, and professional meetings.
7. Overall, the most important sources for teachers to develop their pedagogical knowledge in their teaching career are their own teaching experience and reflection and daily exchanges with colleagues. Teachers' inservice training and organized professional activities such as professional meetings also play an important role in improving their pedagogical knowledge and skills.
8. Last, most teachers think teachers lacking enough time and energy because of heavy teaching loads presents the greatest barrier for them to improve their pedagogical knowledge. Some also believe teachers lacking motivation is a major problem to overcome. Most teachers suggest one way or another to promote their daily exchanges with colleagues for the purpose of developing their pedagogical knowledge.

All of the findings discussed in this chapter are consistent with the major findings described in the previous three chapters.

Chapter 9

Conclusions, Implications, and Recommendations

The final chapter of the Chicago study consists of three sections. The first section provides a summary of the study, its main findings, and conclusions. The second describes the implications of the findings of this study for those directly relevant to the course of improving teachers' pedagogical knowledge, including teacher educators, school administrators, and teachers themselves, and suggestions on how to effectively develop teachers' pedagogical knowledge. The third section recommends a number of possible directions for researchers of teacher knowledge to further undertake study in this line.

Summary and Conclusions

The general question of the study is, how do teachers develop their knowledge in the domain of pedagogy? Specifically, the study has two main research questions: are there different sources of teachers' pedagogical knowledge? And, if there are, how do different sources contribute to the development of teachers' pedagogical knowledge?

This study conceptualizes knowledge from the dynamic relation of the knower (the subject), the known (the object), and the knowing (the interaction of the subject and the object). A subject's knowledge of an object is defined as a mental result of certain interactions of the subject and the object. Following this definition, teachers' pedagogical knowledge is defined as what teachers know about how to teach, which consists of three main components: pedagogical curricular knowledge (PCrK), knowledge of instructional materials and resources, including technology; pedagogical content knowledge (PCnK), knowledge of ways to represent mathematics concepts and procedures; and pedagogical instructional knowledge (PIK), knowledge of instructional strategies and classroom organizational models. The study takes into account the whole life of teachers including their learning experience, preservice training, and inservice experience to investigate the sources from which teachers gain their pedagogical knowledge.

The subjects of this study are all the 77 mathematics teachers in three high-performing high schools, a stratified random sample from the 25 top-scoring high schools in the metropolitan area of Chicago on the Illinois Goal Assessment Program mathematics test.

Three instruments were designed and used to collect the data from the research sample. The first is a questionnaire, which was administered to all the subjects and returned by 69 of the teachers, a response rate of nearly 90%. The second is classroom observation, which was designed to identify what pedagogical knowledge teachers possess and utilize in their teaching, and applied to nine teachers with three teachers from each school: one randomly selected from teachers with 0–5 years' teaching experience, one from those with 6–15 years' teaching experience, and the other 16 or more years'. For each teacher, two classes of different mathematics courses were observed. The third is interview, which was conducted with the nine teachers observed and the three math chairs in these schools, and centered on how teachers learned the specific pedagogical knowledge demonstrated in the classes observed and on how they pursue the improvement of their pedagogical knowledge.

Quantitative methods were employed to analyze the data collected from the questionnaire to obtain some general patterns about how teachers develop their pedagogical knowledge. In addition to descriptive statistics such as percentages, means, and standard deviations, log-linear regression models were used to identify the relative importance of different sources to the development of teachers' pedagogical knowledge; and chi-square tests were applied to detect whether teachers' teaching experience affects the relative importance of different sources to the development of their pedagogical knowledge.

Conclusions, Implications, and Recommendations 211

Qualitative methods were used on the data collected from the classroom observation and interviews to examine contextualized evidence and depict how teachers develop the particular pedagogical knowledge demonstrated in the classrooms observed.

The study yields several main findings and conclusions, which provide answers to the two research questions.

To the first question, are there different sources of teachers' pedagogical knowledge? The answer is simply "yes".

In other words, this study reveals that there exist various sources from which teachers develop all the three components of their pedagogical knowledge: PCrK, PCnK, and PIK. They gain and improve their knowledge of how to teach mathematics from their experience as students, their preservice training, their inservice experience, and even from their social and parenting experience.

As concerns the second question, how do different sources contribute to the development of teachers' pedagogical knowledge? The study produces the following conclusions which apply to teachers overall.

First, with respect to PCrK, teachers' "own teaching experience and reflection" and their "informal exchanges with colleagues" are the two most important sources for all the three sub-components of teachers' PCrK: knowledge of textbooks, knowledge of technology, and knowledge of concrete materials. "Inservice training" is one of the most important sources for teachers' knowledge of technology, and a secondarily important source for their knowledge of teaching materials and concrete materials. "Organized professional activities" is a secondarily important source for all the three components. "Experience as student" is a secondarily important source for teachers' knowledge of teaching materials, and one of the least important sources for their knowledge of technology and concrete materials. "Reading professional journals and books" and "preservice training" are the least important sources for all the three components of teachers' PCrK.

Statistically, all the sources carry the same importance to teachers with different lengths of teaching experience in developing their knowledge of textbooks. However, sources "experience as student(s)" and "preservice training" have more importance to younger teachers than to senior teachers in developing their knowledge of technology, and "preservice training" has more importance to younger teachers in developing their knowledge of concrete materials than their senior colleagues.

Second, with respect to PCnK, the most important sources are teachers' "own teaching experience and reflection" and their "informal exchanges with colleagues". The secondarily important sources are "experience as student", "organized professional activities", "inservice training", and "reading professional journals and books". The least important source is "preservice training". In addition, "textbooks" are also a secondarily important source of teachers' PCnK.

All of those sources have the same importance to teachers with different lengths of teaching experience, except for the source of "informal exchanges with colleagues", from which younger teachers learned significantly more than their senior colleagues.

Third, with respect to PIK, the most important sources are teachers' "own teaching experience and reflection", "informal exchanges with colleagues", and "inservice training". The secondarily important source is "organized professional activities". And the least important sources are "preservice training", "experience as student", and "reading professional journals and books".

In addition, in developing PIK, each of the different sources carries statistically the same importance to all teachers, regardless of when they began teaching. Table 9.1 presents another way to look at the above main findings.

In short, taking into account all the components of teachers' pedagogical knowledge together, we can see that teachers' "own teaching experience and reflection" and "daily exchanges with colleagues" are the two most important sources for them to develop their pedagogical knowledge; "inservice training" and "organized professional activities" are also relatively important sources; but teachers' "experience as student", "preservice training", and "reading professional journals and books" are the least important sources.

The findings of this study do not support the common assumption "teachers teach the way they were taught". More specifically, for the sample of this study, the statement is not just oversimplified, but basically wrong.

Table 9.1 A summary of the main findings on the relative importance of different sources to the development of teachers' pedagogical knowledge

	PCrK				
Sources[a]	of teaching materials	of technology	of other teaching resources	PCnK[b]	PIK
Most important	G, E	G, E, C	G, E	G, E[c]	G, E, C
Secondarily important	C, D, A	D	D	D, A, C, F	D
Least important	F, B	F, A[c], B[c]	F, B[c], A	B	B, A, F

[a]A = Experience as student(s), B = Preservice training, C = Inservice training, D = Organized professional activities, E = Informal exchanges with colleagues, F = Reading professional journals and books, G = Own teaching experience and reflection.
[b]"Textbooks" are also a secondarily important source.
[c]These sources are significantly more important to younger teachers than to senior teachers.

Implications for Teacher Educators, School Administrators, and Teachers

This study is about teachers' knowledge development, a crucial component of teachers' professional development. The findings of this study have several implications for teacher educators, school administrators, and teachers themselves, who all play a role in the development of teachers' knowledge.

For teacher educators, the findings of this study are (at least to this researcher as a former and future teacher educator) disappointing, yet also challenging. Although there are a small number of teachers who consider their preservice training very useful or useful, teachers overall (whether they are young or old) regard their preservice training as the least important source for them to develop their pedagogical knowledge in all the aspects (PCrK, PCnK, and PIK), and a large portion of teachers did not receive necessary training in some basic pedagogical knowledge and skills. Within the curriculum structure of preservice teacher education, student teaching is believed by teachers to be very useful in developing their pedagogical knowledge, but both general educational courses and mathematics methods courses are believed to be much less useful.

According to the study, it will be necessary for teacher educators to take some substantial steps to reform preservice training programs and improve their overall quality. In particular, student teaching should be strengthened and given more time, and the educational and methods courses need to be updated quickly so that more and newer developed pedagogical knowledge and skills are integrated into the courses and introduced to prospective teachers.

For school administrators, the most important message this study reveals is that it is very important and beneficial to promote teachers' daily exchanges with their colleagues in seeking teachers' professional development, since it is one of the most important sources for them to enhance their pedagogical knowledge. From this perspective, some measures are helpful as well as not very difficult to be taken, e.g., strengthening mentoring programs for new teachers and having one or a few offices for all mathematics teachers.[1]

Inviting knowledgeable speakers to schools, supporting teachers to pursue further professional training, and letting them attend outside (especially at the national/regional level) professional activities at least once a year, are also supported by the findings of the study.

Noticing it might raise other difficulties for schools, I think the findings of this study also suggest that they need to pay more attention to motivating teachers and, more importantly, give them reasonable time by reducing their preparation time (teaching less courses if not teaching less classes) and other miscellaneous duties, because collaboration needs time and support (also see Adelman, Haslam, & Pringle, 1996, pp. 78–80).

The study also reveals one necessity of reducing "teacher attrition". Dreeben described teacher attrition as "a persistent time and energy-consuming problem for school administrators" (Dreeben, 1996, p. 109); the findings of this study imply it is also a problem of loss of knowledge, as teachers' own teaching experience and reflection and their daily exchanges with colleagues are the two most important sources for them to develop their pedagogical knowledge, and one cannot expect the sources to be available to new recruits until they begin to assume teaching in

---

[1] It happened that one school I observed in this study assigned each mathematics teacher a classroom which also served as the teacher's office. I believe it is a disadvantage to teachers' regular exchanges with their other colleagues from the viewpoint of this study, though the school might have other reasons for doing so.

schools. Nonetheless, the study does not provide the solutions to the problem.[2]

For teachers, the findings of this study explain that teachers are life-long learners of how to do their teaching job more effectively. In seeking to develop their pedagogical knowledge, they need to be reflective, accumulative, associative, and attentive.

To be reflective means teachers need to reflect on their own teaching practices intentionally and carefully, for example, think about what works, what does not work, why it works or does not work, how to improve it, and so forth, as it is found in this study that teachers' own teaching experience and reflection are the most important source for teachers to develop their pedagogical knowledge. Many researchers have argued and documented the importance of a practitioner's reflection on their experience (e.g., see Schön, 1983; Russell & Munby, 1991). This study presents new empirical evidence which is basically consistent with these researchers' proposals.

To be accumulative means teachers need to gradually establish a store of pedagogical knowledge, which is especially important in developing their PCnK — the ways to represent mathematics topics. This study reveals, without intentional accumulation either in memory or in more written forms, teachers do not necessarily increase their pedagogical knowledge as their experiences grow, which is particularly evident in the interview reported in Chapter 6.

By "to be associative", I specifically mean that teachers need to exchange ideas with their colleagues regularly, as this study finds that kind of activity is one of the most important ways for them to develop their knowledge of pedagogy, and teachers can acquire knowledge of textbooks, technology, mathematics presentations, and general teaching strategies, and so on, from such exchanges.

To be attentive means teachers need to have open minds and be active learners, as the study proves that there are many different sources by which teachers can develop their pedagogical knowledge. Even though overall teachers learn most from their own teaching experience and reflection, and

---

[2]For a more detailed discussion of the problem, see Dreeben (1996) and the National Center for Education Statistics (1994, 1997).

daily exchanges with colleagues, some teachers also learn a considerable amount of pedagogical knowledge from inservice training, professional meetings, and relevant journals and books. How much one can learn from various sources largely depends on how much he/she attends to the sources and effectively uses them.

Recommendations for Further Study

This study focuses on the issue of how teachers develop their knowledge in the domain of pedagogy. It is natural as well as interesting to ask similar questions in other domains of teachers' knowledge.

For example, how do teachers develop their subject knowledge? Can we apply the conclusions of this study to this issue? To me, the answer seems to be "no", as preservice training is supposed to be one of the most important sources for teachers to develop their subject knowledge. Nonetheless, lacking further conceptualization and certain empirical evidence, the answer is largely a common assumption, by no means guaranteed. Moreover, we see in the literature review of Chapter 2 that researchers have proposed that teachers need other kinds of knowledge, such as knowledge of students, knowledge of educational philosophy, and even knowledge of self. The question always exists that, if teachers are expected to have a certain kind of knowledge, then how can they develop it, or how can others, such as teacher educators, help them to develop it?

Although the subjects of this study are mathematics teachers from three of the best-performing high schools in the metropolitan Chicago area, I believe the results of this study can also be reasonably applied to mathematics teachers in the best high schools in other metropolitan areas, or even in towns or rural areas. However, what would be the findings if the subjects were from general (average) high schools or from relatively poor urban schools? As there are so many general and poorer schools, the importance of conducting such studies in those schools is evident.

Of course, the study can be extended to teachers of other school subjects, and also to teachers at elementary and middle-school levels.

I have pointed out that, based on my experience of learning and teaching in China, experience as students should be a major source of teachers' knowledge of textbooks, but this is not the case according to the findings of this study (see Chapter 5). During the study, I was also surprised to find that "reading professional journals and books" was one of the least important sources for teachers in this study to develop their pedagogical knowledge; my personal experience in China leads me to think this is probably not the case for teachers in China because of the differences in school and cultural environments between the two countries. However, are there other major differences between different countries on this issue, and if there are, how can the differences enlighten our understanding of the problem and help to find possible solutions? Being aware of the potential difficulties and the fact that most available international comparisons have been on students, I believe it would be very interesting as well as valuable to undertake international comparative studies on how teachers develop their knowledge.

Part II

The Singapore Study

Chapter 10

The Singapore Study

The second part of this book is comprised of two chapters. Chapter 10 presents an overall report of the Singapore study. Chapter 11 juxtaposes the Singapore study with the Chicago study and provides a comparison of the two studies with focus on the similarities and differences of the findings. Some concluding remarks are also provided at the end of Chapter 11.

The report below about the Singapore study is comprehensive in scope so as to maintain the necessary integrity and completeness of the study. Nevertheless, it is also brief in many aspects, considering the fact that the Singapore study is largely a replication of the Chicago study and many elements such as research questions, conceptual frameworks, research instruments, and methods for data collection and analysis between the two studies are similar, if not the same. Accordingly, the description and discussion below are focused more on the context of the Singapore study and the differences between the two studies, and repetition or recapitulation is kept to a minimum.

Background and Introduction of the Singapore Study

In order to better understand the Singapore study, we shall start with a brief description about Singapore's social and cultural background.

Singapore is a multiracial and multicultural city country, located in Southeast Asia with an area of about 700 square kilometers and a population of about 5 million, including Singapore citizens and permanent residents. According to the latest National Census of Population in 2010, the majority (74.1%) of Singaporean citizens are Chinese, 13.4% are Malays, 9.2%

are Indians, and 3.3% are others (Eurasians and other groups).[1] Chinese, Malay, Tamil, and English are all official languages, but English is the working language and the medium of instruction in all schools.

Historically, Singapore was a British colony for a long time between the beginning of the modern history of Singapore in 1819 and becoming an independent country in 1965. Since then, Singapore has transformed into a highly developed country, which provides favorable conditions for the government and society to offer strong financial support for its educational development including teacher recruitment, training, and professional development. The International Monetary Fund's World Economic Outlook Database (2013) shows that, in terms of gross domestic product (GDP) based on purchasing-power-parity per capita, Singapore is the third richest country with its GDP per capita at US$61,567. In comparison, the US is ranked seventh with US$51,248.[2]

Like many other Asian countries, Singapore has a centralized national education system. Starting from age six, children usually receive six years of compulsory primary education from the first grade in primary school, which is more commonly known as Primary 1 or simply P1 in Singapore, to Primary 6 or P6. At the end of primary education, all students take the Primary School Leaving Examination (PSLE), a national high-stake test, in the four core subjects, mathematics, science, English, and Mother Tongue.

According to their PSLE results, students will then be streamed into one of the different courses in secondary education: Special Course, Express Course, and Normal Course, which is further spilt into Normal Academic Course and Normal Technical Course. The streaming policy is designed to cater to students' different learning abilities, interests, and future needs. Students in Special or Express Courses follow a four-year program, at the end of which they will take the Singapore–Cambridge General Certificate of Education "Ordinary" Level Examination (GCE, "O" Level). Students in Normal Academic and Normal Technical Courses follow a four-year program leading up to GCE "Normal" Level Examination ("N" Level), with the possibility for better students to study for a fifth year to take the GCE "O" Level Examination. In 2004, the Ministry of Education (MOE)

---

[1] Source: Singapore Department of Statistics (http://www.singstat.gov.sg/).
[2] Source: International Monetary Fund (http://www.imf.org/).

announced that selected students in the Normal Course would have the opportunity to sit for the GCE "O" Level Examination directly without first taking the GCE "N" Level Examination.

After secondary education and based on their results in GCE "O" or "N" Level Examinations, students can progress to their post-secondary education in the two-year junior colleges or a three-year centralized institute to study a pre-university course, which leads to the GCE "Advanced" Level Examination ("A" Level). Another option is the three-year polytechnics, which offer a large variety of practice and employment-oriented diploma courses to prepare for knowledgeable middle-level professionals, or the Institutes of Technical Education (ITE) which are mainly provided for producing a skilled workforce. According to their results of post-secondary assessment examinations and their own interests, students may continue to undergraduate studies at university level or join the workforce.

Over the last two decades or so, Singapore's education has received considerable attention from policy makers, education administrators, researchers, and practitioners in many different countries. This is closely related to the fact that Singapore students have consistently demonstrated stellar performances in highly visible and well-referred international comparative studies, such as the Third International Mathematics and Science Study, which was later renamed the Trends in International Mathematics and Science Study (TIMSS) and has been conducted on a regular four-year cycle since 1995, assessing students' performance in both mathematics and sciences in different grade levels, and the Programme for International Student Assessment (PISA) which takes place on three-year cycles starting from 2000 and measures 15-year-old school students' scholastic performance in mathematics, sciences, and reading.[3]

Table 10.1 shows the rankings of Singaporean students' performances in the TIMSS studies from 1995 to 2011 in terms of their average scores and the number of participating countries and education systems (n).

---

[3] Singapore only participated in the PISA study in 2009 and 2012. In PISA 2009, Singapore students' performance, in terms of the mean score, was ranked second in mathematics, fourth in science, and fifth in reading, while in PISA 2012, Singapore students' performance was ranked second in mathematics and third in both science and reading. (Source: http://www.oecd.org/pisa/).

Table 10.1 Singaporean students' performances in TIMSS mathematics studies from 1995 to 2011

	1995	1999	2003	2007	2011
Grade 4	1st (n = 25)	—	1st (n = 25)	2nd (n = 36)	1st (n = 52)
Grade 8	1st (n = 41)	1st (n = 38)	1st (n = 46)	3rd (n = 49)	2nd (n = 45)

Note: The TIMSS results are widely available on-line to the public, e.g., see http://timssandpirls.bc.edu/.

In searching for the reasons behind the successful story of Singapore's acclaimed education system, many researchers have listed Singaporean teachers' quality and work ethos, initial teacher training, and teacher professional development as the first or key contributing factors (e.g., Ministry of Education, 1995). As Sing Kong Lee, Director of the National Institute of Education, which is the sole institute of teacher training in the nation, said, "at the heart of Singapore's educational success is the quality of our teachers" (Lee, 2012, p. 30).

The Singapore study described hereafter is targeted at the issue of "teacher professional development" or "teacher learning" in the context of the Singaporean education system.[4] It aims to examine how teachers develop their knowledge in the domain of pedagogy. The general research question is: how do Singapore teachers develop their pedagogical knowledge?

More specifically, taking all the mathematics teachers in a randomly selected sample of six secondary schools in Singapore as the research subjects, the study is intended to address the following two related research questions:

1. Are there different sources of Singaporean teachers' pedagogical knowledge?

---

[4]Part of the chapter below was presented at the 2005 American Educational Research Association Annual Meeting, Montreal, Canada (Fan & Cheong, 2005). Significant changes were made for this book.

2. If the answer to the first question is "yes", then how do different sources contribute to the development of Singaporean teachers' pedagogical knowledge? In other words, if there are different sources, what is the relative importance of these sources to the growth of Singaporean teachers' pedagogical knowledge?

Through investigating the two specific research questions, it is hoped that the study can obtain a relatively systematic picture of how mathematics teachers in Singapore develop their knowledge of pedagogy, and hence explore its implication for teachers' learning and professional development and shed light on how to effectively improve teachers' pedagogical knowledge.

The Singapore study can be considered, on the one hand, as a replicated study of the Chicago study in many aspects. On the other hand, the study is more than just an extension of the Chicago study. It has its own theoretical as well as practical purposes. There are mainly three reasons that it aims to investigate how teachers develop their pedagogical knowledge in the Singaporean educational context.

First, because the US and Singapore have very different teacher education systems, social and cultural traditions, and school environments, what is true in Chicago is not necessarily true in Singapore. It is hoped the Singapore study will help verify which findings from other studies have general meaning for teachers' knowledge development, and which do not, and hence advance the scholarship in this area.

Second, as the research subjects in the Singapore study are mathematics teachers in Singaporean schools, the study will have practical values and implications for mathematics teachers, school administrators, the teacher educators, and policy makers in Singapore in their pursuing teachers' professional development.

Third, no matter whether or not the results are consistent between the two studies, the comparison between the US and Singapore can lead us to explore the similarities and differences concerning the issues of teacher learning, and deepen our understanding of how teachers develop their pedagogical knowledge in different social and educational contexts. As Fischer argued when he compared educational mathematics between the East and the West, conformities or similarities can enable us to understand each other, while differences can present us with the chance to enrich and to complete each other (Fischer, 2006).

The Conceptual Framework

The conceptual framework about knowledge, mathematics teachers' pedagogical knowledge, and its sources established in the Chicago study was basically adopted in the Singapore study. Hence, in some parts below, the description is brief and concise, as mentioned at the beginning of this chapter.

Knowledge

Epistemologically, knowledge was conceptualized from the dynamic relation of the knower (the subject of knowledge), the known (the object of knowledge), and the knowing (the interaction of the subject and the object). A subject's knowledge of an object was thus defined as a mental result of certain interactions of the subject and the object.

In the definition, the subject means teachers, the object means something about pedagogy, a mental result means the subject's cognitive achievement including understanding, memory, or belief, and certain interactions include the subject's acquainting, observing, experiencing, reflecting, reasoning, thinking, and other similar actions against the object.

It should be pointed out that the purpose of the study is to investigate how teachers develop their pedagogical knowledge, but not to evaluate how much pedagogical knowledge teachers have, nor to assess if the pedagogical knowledge they possess is correct or wrong, desirable or undesirable. Moreover, according to the above definition, we take into account all kinds of knowledge as long as it is a mental result developed from some interaction, no matter whether it is explicit or implicit (tacit), direct or indirect, theoretical or practical, about "that" or about "how", and so on.

Pedagogical knowledge

The Singapore study also largely employed the concept about teachers' pedagogical knowledge by the National Council of Teachers of

Mathematics (NCTM, 1991) and categorized pedagogical knowledge into three components:

1. Pedagogical Curricular Knowledge (PCrK) — knowledge of teaching materials and resources, including technology.

Examples of teachers' PCrK include their knowledge about the syllabus, textbooks, workbooks, and other teaching materials, technological tools (most importantly calculators and computers/software), and concrete materials or physical equipment for teaching mathematics, and so on.

2. Pedagogical Content Knowledge (PCnK) — knowledge of ways to represent mathematics concepts and procedures.

Taking the topic of quadratic equations as an example, PCnK deals with questions like the following: how to introduce the relevant concepts and skills in this topic to students with different academic backgrounds? What is the easiest or most difficult part in learning this topic? How to make the topic more interesting or challenging to different students? What are possible models of representing the concepts or procedures to students, and what are the advantages and disadvantages of those models (also see Marks, 1990; Cochran, DeRuiter, & King, 1993)?

3. Pedagogical Instructional Knowledge (PIK) — knowledge of general teaching strategies and classroom organizational models.

Teachers' PIK is not particularly about teaching a specific topic in mathematics, and in a large sense, it is not even particularly about teaching a specific school subject including mathematics. It generally deals with questions like how to organize classrooms, how to motivate and interact with students, how to pose questions to students and answer their questions, how to assess students' learning, and so on. Examples of PIK include teachers' knowledge about cooperative learning, constructivism learning, and other general teaching techniques and organizational models (also see Johnson, 1982, 1986; Jones & Vesilind, 1996).

Sources of pedagogical knowledge

To examine the possible sources from which teachers could develop their pedagogical knowledge, the Singapore study also took into account the whole life of teachers, including (a) their learning experiences mainly in formal educational environments (school) but also in informal educational environments before accepting preservice training, (b) their preservice training experiences, and (c) their inservice experiences.

Figure 10.1 shows the framework consisting of the above three main components and several sub-components adapted from the Chicago study and used for investigating the sources of teachers' pedagogical knowledge in the Singapore study.

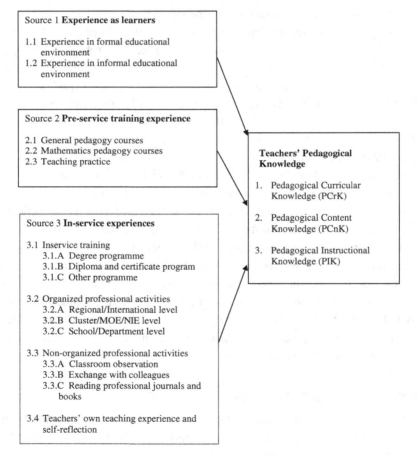

Figure 10.1 A framework to investigate the sources of Singaporean teachers' pedagogical knowledge

Particular attention was paid to the following seven sources in the study, which were conceived as the major sources for our investigation, especially for the questionnaire survey: (1) experience as (school) students, (2) preservice training, (3) inservice training, (4) organized professional activities, (5) informal exchanges with colleagues, (6) reading professional journals and books, and (7) teachers' own teaching experience and reflection. Nevertheless, other possible sources also received necessary attention during the study, which was particularly evident during the interviews with teachers for open-ended questions after the classroom observation.

It should be pointed out that Singapore's educational context and teacher development practices including terminology have been taken into account and hence reflected in the framework shown in Figure 10.1.

For example, in Source 3: Inservice experiences, "3.1.B Diploma and certificate program" was added because in Singapore inservice training that leads to diplomas and certificates has been offered for a long time. In fact, in the early days many school teachers, especially primary school teachers, held diplomas or certificates in education, which were commonly acceptable qualifications for being a teacher in Singapore, though the number of these teachers has declined over the last two decades because of the fast development of the economy and other aspects of the society. Currently, the National Institute of Education still offers a small number of advanced diploma, diploma, and certificate programs for inservice teachers to cater for the needs of the education system.

In relation to "Source 3.2 Organized professional activities", it should be pointed out that in Singapore the word "region" generally refers to the area of Southeast Asia, which is different from the US where the term usually refers to an area of a few neighboring states within the same country. "Cluster" is an intermediate level above the school level in the education organizational structure, introduced in the late 1990s. The schools are grouped, mainly based on geography, into 29 clusters and each cluster is facilitated by a Cluster Superintendent.[5] As stated earlier, Singapore is a small city country, and unlike the US, does not have county and state divisions.

---

[5] See http://sis.moe.gov.sg/SchoolClusters.aspx.

Methodological Matters

Population and sample

The school population in the study consisted of all of the 152 secondary schools in Singapore. The study sample consisted of six schools, a stratified random sample from the population. One school was randomly selected from the top 25 schools, one from the next best 25, and the remaining four from the rest of the population, according to the school ranking by the Singaporean MOE which was based on their students' average performance in the GCE "O" Level Examination in 1998. The ranking has been rather stable over the years.

All the 73 mathematics teachers in the sample schools participated in the first stage of this study, that is, a questionnaire survey. At a later stage, 18 of them were observed for classroom teaching, and 22 of them were further interviewed.

In order to keep those six participating schools anonymous, they will be hereafter labelled Schools 2A, 2B, 2C, 2D, 2E, and 2F respectively. Table 10.2 shows the numbers of students, teachers, and mathematics teachers in these schools.

Other information such as the distribution of teachers in terms of gender, age, and teaching experience across the participating schools can be found in Appendix 2A.

Instruments

Three instruments were employed to collect the original data in the study: a questionnaire, classroom observation, and teacher interview. The

Table 10.2 Numbers of students, teachers, and mathematics teachers in the six Singaporean schools

Number of	School						Total
	2A	2B	2C	2D	2E	2F	
Students	1,202	895	1,200	1,500	1,340	1,090	7,227
Teachers	65	52	60	70	65	55	367
Mathematics teachers*	9	10	9	15	16	7	66

*The figures do not include those who did not return the questionnaire in the study.

instruments were adopted from the Chicago study with necessary adaptation and modification so as to better fit the Singaporean educational context and social background.

The questionnaire was based on the conceptual framework of the study and comprised 23 questions. The items designed in the questionnaire were basically targeted at the collection of three sets of information: on teachers' background, on how teachers' specific experiences contributed to the development of their pedagogical knowledge, and on how teachers' specific pedagogical knowledge was developed from their different experiences (see Appendix 2B).

The questionnaire was distributed to all the 73 teachers and collected from 66 of them, a response rate of 90.4%. Among the 66 teachers, 33 teachers had less than six years of teaching experience, 13 teachers had 6–15 years, and 20 teachers had 16 or more years.

Classroom observation was designed to identify what pedagogical knowledge teachers would demonstrate in their actual teaching practice, so appropriate questions could be raised during the following interviews to find out the sources from which teachers developed the relevant pedagogical knowledge. For this purpose, the classroom observation focused on the pedagogical aspects, that is, what and how the teachers used the teaching resources (including textbooks, concrete materials, computers, and calculators) to represent mathematical concepts or procedures to their students, and how they employed teaching activities and organized the classrooms. As mentioned earlier in the Chicago study, classroom observation is necessary because teachers' pedagogical knowledge can often be "tacit" and "contextualized", in which case observation is a more feasible and reliable way to detect it.

In total, 18 teachers, three from each school, were selected for classroom observation. In each participating school, one teacher was randomly selected from the first group of teachers who had taught mathematics for less than six years, another from the second group who had taught for 6 to 15 years, and the other from the third group with 16 or more years of teaching experience. The stratified random sample was selected to detect the possible differences of the sources of their pedagogical knowledge for teachers with different teaching experience.

Table 10.3 displays relevant information about the 18 teachers observed. Each teacher in the sub-sample was observed for two consecutive

Table 10.3 Selection of teachers for classroom observation in the Singapore study

School	0–5 years	6–15 years	16+ years	Average years of teaching experience
School 2A Teacher selected	STA1 (3 years)	STA2 (10 years)	STA3 (29 years)	14 years
School 2B Teacher selected	STB1 (3 years)	STB2 (15 years)	STB3 (34 years)	17 years
School 2C Teacher selected	STC1 (2 years)	STC2 (6 years)	STC3 (24 years)	11 years
School 2D Teacher selected	STD1 (3 years)	STD2 (6 years)	STD3 (39 years)	16 years
School 2E Teacher selected	STE1 (3 years)	STE2 (10 years)	STE3 (16 years)	10 years
School 2F Teacher selected	STF1 (1 year)	STF2 (7 years)	STF3 (16 years)	8 years
Total	6	6	6	10 years

Note: The figures in brackets are the numbers of years of teaching experience.

class periods of about 60–70 minutes. For the reliability of data collected and the representativeness of classrooms observed, teachers were informed that normal classes were to be observed and no special preparations were required.

All classroom observations were carried out between April and October 2000, and a common guide, directly taken from the Chicago study (see Appendix 1B), was used for classroom observation. All the classroom observations were documented using audio recorders and field notes. Table 10.4 lists the information about teachers and topics of the classes observed. It shows clearly that a wide range of contents were covered in the classes observed.

Interviews were conducted with 18 teachers (including two heads of departments (HODs)) whose classroom teaching was observed and four HODs without involving classroom observation. As in the Chicago study, the purpose of the interview was not to make a comparison of the importance of different sources for teachers to develop their pedagogical knowledge. Instead, it was designed to obtain an in-depth and contextualized understanding of the sources of teachers' specific pedagogical knowledge.

Table 10.4 Classes observed in the Singapore study

School	Teachers	Classes observed
2A	STA1	Simple trigonometric identities and equations
	STA2	Basic geometrical ideas and properties: angle, lines
	STA3	Pythagoras theorem
2B	STB1	Mean, median and mode
	STB2	Review for "O" level test
	STB3	Review on graphs of functions
2C	STC1	Measures: mass and money
	STC2	Vectors
	STC3	Obtuse angle, sine–cosine ratio
2D	STD1	Common logarithms
	STD2	Quadratic graphs
	STD3	Parametric equations
2E	STE1	Logarithm and coordinate geometry (review)
	STE2	Problem solving
	STE3	Algebraic expression and factorization
2F	STF1	Area and perimeter of composite figures
	STF2	Interest and percentage
	STF3	Properties of quadrilateral (Venn diagram)

The interview with each of the 18 teachers (including two HODs as teachers) was conducted after the classroom observation. The interview focused on how they learned the specific pedagogical knowledge demonstrated in the classes observed. Meanwhile, the interviews with all the HODs focused more on the professional school environment for mathematics teachers to develop their pedagogical knowledge. The guidelines designed in the Chicago study, which can be found in Appendices 1C and 1D, were also employed to guide the interview with teachers and HODs, respectively, in this study. Nevertheless, it should be noted that specific questions for the interview would be contextualized, depending on the actual classes observed (for teachers) and schools visited (for HODs). Each interview lasted approximately 45 minutes and was audiorecorded and later transcribed for data analysis.

It should also be emphasized that all the instruments described above were not intended to detect how much knowledge the participating teachers had or to measure how good their knowledge was, which was beyond the purpose of the study. Even though there were such questions included in the questionnaire (for example Q19a and Q19b) and they were asked during the actual interviews, their purpose was mainly to provide relevant information needed for the researchers to understand and analyze the teachers' responses better, and sometimes to serve as stimulants for teachers to recall their experiences in order to answer other questions that were directly targeted at how teachers develop their pedagogical knowledge, the purpose of the study.

Data analysis

Like in the Chicago study, both quantitative and qualitative methods were employed to analyze the data collected. Quantitative methods were mainly utilized to analyze the data collected from the questionnaire survey. Descriptive statistics such as mean, standard deviation, and percentage were applied to describe the general pattern of how different sources contributed to the development of teachers' pedagogical knowledge. Logistic regression analysis was used to detect whether there existed significant differences between the contributions of different sources to teachers' pedagogical knowledge. Chi-square tests were used to examine the influence of teachers' different backgrounds (in terms of the three groups mentioned above) on the development of their pedagogical knowledge.

Qualitative methods were mainly used to analyze the data collected from classroom observations and interviews to depict in-depth how individual teachers developed their specific pedagogical knowledge. A coding system for data analysis was established so that the teachers' pedagogical knowledge was categorized into PCrK, PCnK, and PIK, and the sources of a teacher's pedagogical knowledge were first classified into his/her experience as a learner, preservice experience, and inservice experience; then sub-coded into more specific sources like his/her experience as a student, preservice training, inservice training, (attending) organized professional activities, informal (daily) exchanges with colleagues, reading professional journals and books, and their own teaching experience (practices) and reflection. In the initial reading of the questionnaire and the transcription of the interviews, if the teachers' responses (mainly to Questions 21 and 22, which are open-ended) could not be accounted for by the established coding system, then another category would be created and explained.

It should be noted that the data source for the study was limited to secondary school mathematics teachers in the Singaporean educational system. Moreover, the results of the study were mainly based on the teachers' responses to the questionnaire and interview. Therefore, some caution should be exerted in interpreting and generalizing the findings.

Findings of the Singapore Study

As indicated earlier, the Singapore study also classified teachers' pedagogical knowledge into three core components: PCrK, PCnK, and PIK. Therefore, we shall report the results of the study based on the three components of teachers' pedagogical knowledge.

Pedagogical curricular knowledge

As mentioned earlier, PCrK is about teaching materials and resources. According to NCTM (1991), teachers often rely on a variety of instructional materials and resources, including textbooks, booklets, calculators, computer software, and so on to conduct classroom teaching. For this purpose, they need to identify, use, and assess these resources effectively in their instruction. In the study, teachers' PCrK was mainly investigated through their knowledge of teaching materials (primarily textbooks), of technology (primarily calculators and computers/software), and of other teaching resources (primarily concrete materials).

Knowledge of textbooks

Researches around the world have generally and consistently revealed that textbooks overall play a dominant role in instructional practice in mathematics classrooms (e.g., Netherlands: Krammer, 1985; Germany: Bierhoff, 1996; USA: Woodward & Elliott, 1990; Usiskin, 2013; Philippines: Ulep, 2000; Japan: Fujii, 2001; Singapore: Zhu & Fan, 2002; China: Fan *et al.*, 2004; also see Fan, Zhu, & Miao, 2013). However, as stated earlier, how much textbooks really affect teachers' teaching to a large extent depends on how much knowledge the teachers have about the textbooks they are using for teaching mathematics, which further depends on how they got to know about the textbooks.

The data from the questionnaire survey revealed that more than 57% of the Singapore teachers employed "students' reading the texts" as one teaching strategy and many more teachers (nearly 68%) noted that their teachers also asked them to do so when they were school students. Furthermore, from teachers' responses to Question 19b, which was particularly targeted on teachers' knowledge of textbooks in the questionnaire, nearly 95% of the teachers felt that their knowledge of textbooks was either "fairly good" or "very good". It is interesting to note that in the Chicago study, the percentage is 99%; both studies showed teachers were confident about their knowledge of textbooks used in their teaching.

Based on the responses of teachers to Question 19c, Table 10.5 tabulates the detailed distribution of the percentage of teachers giving different

Table 10.5 Results on how Singaporean teachers gave different evaluations on the contribution of various sources to the development of their knowledge of textbooks

Sources (Number of teachers)	Contribution				Average evaluation*
	Very much	Somewhat	Little	No contribution	
Experience as student (57)	8 (14.04%)	12 (21.05%)	12 (21.05%)	25 (43.86%)	2.05
Preservice training (55)	2 (3.63%)	21 (38.18%)	13 (23.64%)	19 (34.55%)	2.11
Inservice training (55)	4 (7.27%)	20 (36.36%)	12 (21.82%)	19 (34.55%)	2.16
Organized professional activities (55)	2 (3.63%)	18 (32.73%)	17 (30.91%)	18 (32.73%)	2.07
Informal exchanges with colleagues (56)	7 (12.50%)	29 (51.79%)	10 (17.86%)	10 (17.86%)	2.59
Reading professional journals and books (54)	3 (5.56%)	12 (22.22%)	16 (29.63%)	23 (42.59%)	1.91
Own teaching experience and reflection (56)	20 (35.71%)	29 (51.79%)	4 (7.14%)	3 (5.36%)	3.18

*The average evaluation is obtained by using the ordinal scale, 4 = very much, 3 = somewhat, 2 = little, and 1 = no contribution.

evaluations across the four choices concerning how the seven possible sources contributed to the development of their knowledge of the textbooks they were using for mathematics teaching. Those teachers' average evaluation about the contribution of each source is also presented.

From Table 10.5, it is clear that there are a wide variety of sources from which teachers can develop their knowledge of textbooks. Moreover, in terms of those teachers' average evaluation on each source, which was included in the table, the sources from the most important to the least important for them to develop their knowledge of the textbooks they were using during the survey were their "own teaching experience and reflection" (3.18), "informal exchanges with colleagues" (2.59), "inservice training" (2.16), "preservice training" (2.11), "organized professional activities" (2.07), "experience as student" (2.05), and finally "reading professional journals and books" (1.91).

After the above descriptive analysis, logistic regression analysis was further applied to the data using the SPSS software. The results, as shown in Table 10.6, indicate that the order of the importance of the different sources based on the logit models, in terms of the values of the estimates presented

Table 10.6 Results of logistic regression on the data about the contribution of different sources to the Singaporean teachers' knowledge of textbooks

		Estimate	Std. Error	Wald	df	Sig.	95% Confidence Interval	
							Lower Bound	Upper Bound
Threshold	[CONTRIBU = 1]	−3.095	1.382	5.016	1	.025	−5.804	−.386
	[CONTRIBU = 2]	−2.070	1.379	2.254	1	.133	−4.772	.632
	[CONTRIBU = 3]	.197	1.370	.021	1	.885	−2.488	2.883
Location	[STUDEN = 0]	.198	.345	.329	1	.566	−.479	.875
	[INSVC = 0]	−8.575E-02	.347	.061	1	.805	−.767	.595
	[ORGACT = 0]	6.615E-02	.348	.036	1	.849	−.615	.748
	[COLLEAG = 0]	−.902	.351	6.594	1	.010	−1.590	−.213
	[READING = 0]	.403	.352	1.312	1	.252	−.287	1.093
	[OWNREFL = 0]	−2.149	.373	33.238	1	.000	−2.879	−1.418
	[PRESVC = 0][a]	0	.	.	0	.	.	.

Link function: Logit.
[a]This parameter is set to zero because it is redundant.

in Table 10.6, is the same as revealed above using teachers' average evaluation.[6]

Moreover, from each p-value in Table 10.6 in the column "Sig.", which indicates the significant level of the Wald chi-square test statistic,[7] the square of the ratio of the parameter estimate to its standard error, it is clear that compared to "preservice training" [PRESVC], teachers' "own teaching experience and reflection" [OWNREFL] and "informal exchanges with colleagues" [COLLEAG] are significantly more important at the 0.05 level, and "inservice training" [INSVC], "organized professional activities" [ORGACT], "experience as student" [STUDEN], and "reading professional journals and books" [READING] have the same importance as "preservice training".

To find if the length of teaching experience would affect teachers' evaluation of the importance of each source to the development of their knowledge of textbooks, all the teachers were classified into three groups: STG1 consisting of teachers with less than 6 years' teaching experience, STG2 with 6–15 years', and STG3 with 16 or more years'.

Table 10.7 shows the distributions of the numbers of teachers among the three groups giving different evaluations of the contribution of their experience as school students to their knowledge of textbooks.

Applying the chi-square test to the distribution across the three groups, the results indicate there is no significant difference at the 0.05 level ($\chi^2 = 9.601$, df $= 6$, p $= 0.1425$). Similarly, no significant differences were found for the other sources across the three groups. In other words, the length of teachers' teaching experience does not significantly affect their evaluation about the importance of the different sources in developing their knowledge of textbooks.

Different from the questionnaire, we asked the teachers to elaborate the knowledge sources by themselves in the interview. As a result, three of

---

[6] Please note regarding the results of logit regression models: the interpretation offered for Table 1F.1 in the appendix, obtained using SAS, applies to the outputs shown in Table 10.6, as well as in Tables 10.10, 10.13 and 10.20, except that "Threshold" here in the SPSS outputs represents the intercepts and "Location" represents the predictor variables of the logit models.

[7] Note that the p-value detects if the parameter equals zero, that is, the corresponding explanatory variable does not affect the probability predicted using the model.

Table 10.7 Distributions of the numbers of the Singaporean teachers among the three groups giving different evaluations of the contribution of the source "experience as school students" to their knowledge of textbooks

Evaluation	Length of teaching experience			Total
	TG1 (0–5 years)	TG2 (6–15 years)	TG3 (16+ years)	
Very much	5	0	3	8
Somewhat	7	1	4	12
Little	4	6	2	12
No contribution	14	5	6	25
Total	30	12	15	57

the seven sources did not appear in all the interviews at all: "preservice training", "inservice training", and "reading professional journals and books". The finding is somehow consistent with the facts revealed in Moulton's (1997) review that formal trainings on textbooks and their usage were actually insufficient around the world. As to professional reading, many teachers in the interview commented that they had a heavy workload every day so that they could hardly find time to do such reading, although some of them pointed out that professional reading might benefit their teaching practice.

Similar to the findings from the questionnaire survey, the interviews also reveal that teachers' "own teaching experience and reflection" plays the most important role in developing teachers' knowledge of textbooks. Moreover, out of the five teachers who cited this source, four have taught mathematics for more than five years. In contrast, all the three teachers who believed that their past experience as school students helped them to develop the knowledge of textbooks have no more than five years' teaching experience.

The questionnaire survey shows that teachers' "informal exchanges with colleagues" is the second most important source in the development of teachers' knowledge of textbooks. Nevertheless, only two teachers in the interview noted that part of their knowledge of textbooks could be attributed to such exchanging. In addition, one teacher mentioned that attending organized professional activities was also a helpful source for him to develop knowledge of textbooks. However, from the teacher's further description,

Table 10.8 Sources of Singaporean teachers' knowledge of textbooks identified during the interview

	A	B	C	D	E	F	G	Other sources
Frequency of teachers identifying the source	6	0	1	8	2	0	12	7

Note: A = Experience as student, B = Preservice training, C = Inservice training, D = Organized professional activities, E = Informal exchanges with colleagues, F = Reading professional journals and books, G = Own teaching experience and reflection. Other sources: Reading text (6) and Own exploration (1).

we found that to choose a suitable textbook for school use was the only theme of such professional activities.

In the interview, teachers also indicated there were some other sources for them to gain knowledge of textbooks. Many claimed that they acquired much knowledge of textbooks through reading texts in the textbooks themselves. One teacher also mentioned that he got to know certain teaching materials, including textbooks, through publishers' recommendations.

Table 10.8 summarizes the results from the interview. The data clearly revealed there were various sources for teachers to develop their knowledge of textbooks.

Knowledge of technology

Technology was defined in the NCTM *Standards* as one of the instructional resources that teachers need to use in their classroom teaching (NCTM, 1991, p. 151). In Singapore, use of information technology (IT) is one of the three initiatives advocated by the MOE. In fact, the MOE's IT masterplan in education launched in 1997 suggests that students spend up to 30% of curriculum time using IT (Deng & Gopinathan, 2005).

In this study, our main focus of technology was on calculators and computers including relevant software. These technologies have greatly influenced the purposes, the contents, and the ways of teaching and learning mathematics in school for the last two decades (e.g., see Heid, 1997).

The data collected from the questionnaire survey show that there have been notable changes in the use of IT in teaching practice from the past to nowadays, especially in the use of computers. According to teachers' responses to Question 6 in the questionnaire, more than 85% of the teachers had employed "use of computer" as one of their teaching strategies. In contrast, more than 90% of them said that their teachers did not teach them by using computer when they were school students. The discrepancy in the use of calculators between the two time periods is relatively small (past: 76%, now: 100%).

Regarding the frequency of using the two IT tools in teaching, it was found that computers were used much less frequently than calculators. About 57.4% of the teachers reported in the questionnaire survey that they used calculators "always" or "most of the time", and another 31.2% of them sometimes did so. However, there were only 5% of the teachers who used computers in their classrooms "always" or "most of the time", and the majority (68.3%) only did so occasionally. There were some teachers who rarely or never used a calculator (11.5%) or computer (26.7%).

Table 10.9 reports the results from teachers' responses to Question 17b in the questionnaire which requires them to evaluate the contribution of various sources to the development of their knowledge of how to use technology for teaching mathematics. Average evaluation for each source is also included in the table.

According to the results shown in Table 10.9, it is clear there were various sources for teachers to gain their knowledge about technology. Furthermore, according to their average evaluation, the sources from the most important to the least important for them to develop their knowledge of technology were their "own teaching experience and reflection" (3.25), "inservice training" (2.92), "informal exchanges with colleagues" (2.91), "organized professional activities" (2.80), "preservice training" (2.59), "experience as student" (2.14), and "reading professional journals and books" (2.02).

Logistic regression analysis was again used to analyze the data in Table 10.9. The results showed that, in developing teachers' knowledge of technology for mathematics teaching in classrooms, among the seven possible knowledge sources, their "own teaching experience and reflection" and "informal exchanges with colleagues" are significantly more important than "inservice training", "organized professional activities", and "preservice training", which are in turn significantly more important than their "experience as student" and "reading professional journals and books". The detailed statistical results are reported in Table 10.10.

Table 10.9 Results on how Singaporean teachers gave different evaluations on the contribution of various sources to the development of teachers' knowledge of using technology for teaching mathematics

Sources (Number of teachers)	Contribution				Average evaluation[a]
	Very much	Somewhat	Little	No contribution	
Experience as student (64)	6 (9.38%)	20 (31.25%)	15 (23.44%)	23 (35.94%)	2.14
Preservice training (63)	10 (15.87%)	28 (44.44%)	14 (22.22%)	11 (17.46%)	2.59
Inservice training (61)	15 (24.59%)	30 (49.18%)	12 (19.67%)	4 (6.56%)	2.92
Organized professional activities (61)	12 (19.67%)	30 (49.18%)	14 (22.95%)	5 (8.20%)	2.80
Informal exchanges with colleagues (64)	14 (21.88%)	32 (50%)	16 (25%)	2 (3.13%)	2.91
Reading professional journals and books (63)	0 (0%)	20 (31.75%)	24 (38.10%)	19 (30.16%)	2.02
Own teaching experience and reflection (63)	22 (34.92%)	35 (55.56%)	6 (9.52%)	0 (0%)	3.25

Link function: Logit.
[a]The average evaluation is obtained by using the ordinal scale, $4 =$ very much, $3 =$ somewhat, $2 =$ little, and $1 =$ no contribution.

As we can see, the order of importance of the various sources based on the logistic model is the same as that obtained using the previous average evaluations, except for "informal exchanges with colleagues" and "inservice training". Note the results of the logistic model are about the odds ratios, whilst the average evaluations are about the weighted means. Therefore it appears understandable that there exists some difference, and we believe that to a large sense the logistic regression analysis provided us with more accurate information from the data examined since the average evaluation was based on an ordinal scale, as indicated earlier.

Similar to the data on teachers' knowledge of textbooks, in order to find out if there is significant difference among the teachers with different

Table 10.10 Results of logistic regression on the data about the contribution of different sources to Singaporean teachers' knowledge of technology

		Estimate	Std. Error	Wald	df	Sig.	95% Confidence Interval	
							Lower Bound	Upper Bound
Threshold	[CONTRIBU = 1]	−1.915	.950	4.069	1	.044	−3.776	−5.430E-02
	[CONTRIBU = 2]	−.753	.949	.629	1	.428	−2.614	1.108
	[CONTRIBU = 3]	1.460	.947	2.378	1	.123	−.396	3.316
Location	[STUDEN = 0]	.727	.238	9.327	1	.002	.261	1.194
	[PRESVC = 0]	.163	.239	.463	1	.496	−.306	.631
	[INSVC = 0]	−.226	.241	.878	1	.349	−.699	.247
	[COLLEAG = 0]	−.582	.241	5.815	1	.016	−1.055	−.109
	[READING = 0]	.906	.241	14.110	1	.000	.433	1.379
	[OWNREFL = 0]	−1.562	.250	39.027	1	.000	−2.052	−1.072
	[ORGACT = 0][a]	0	.	.	0	.	.	.

Link function: Logit.
[a] This parameter is set to zero because it is redundant.

lengths of teaching experience in their viewing of the contribution of different sources to the development of their knowledge of using technology for teaching mathematics, we employed the chi-square test again to the data, sorted according to the three groups of teachers, STG1, STG2, and STG3 as explained above. The results revealed that there exists no significant difference. In other words, each source overall carries the same importance to those teachers in developing their knowledge of using technology in teaching mathematics, no matter how long they have been in the teaching profession.

The data collected from the interview with 22 teachers show that many teachers who were interviewed expressed that the inservice courses, teaching experience and reflection, and informal exchanges with colleagues were important sources for them to develop their knowledge of how to use the new technology such as calculators and computers in their classrooms.

More specifically, relevant inservice training was mentioned most frequently, in total by ten teachers as a source for them to acquire their IT knowledge. In fact, according to teachers' responses to Question 13 in the questionnaire, compared to the training on textbooks and other teaching materials, more professional training, especially inservice training, on use of IT had been provided to the teachers. The responses suggest that, although only about 26% of the teachers had learnt the use of calculators and 38% of the teachers had learnt the use of computers during their preservice training, more than 82% of the teachers had learnt the use of IT from inservice training during the last five years. Moreover, almost all the teachers (96%) commented that such training was quite useful for their teaching.

Six teachers in the interview also explained that their "own teaching experience and reflection" was very important for them to develop this particular knowledge. Some further commented that although the IT-based professional training was very useful for them to upgrade their IT knowledge, what they had learnt in such training was actually not enough and they must practice more on their own. For instance, STB3 (see Table 10.3) pointed out that attending courses sometimes was not helpful because the organizers did not know exactly what teachers wanted to learn, therefore in this sense, own experiences and reflection became more effective and useful.

As one HOD pointed out, the number of participants for inservice courses was sometimes limited, so some teachers' applications for attending inservice training might be rejected. Therefore, formal and informal sharing from the teachers who had received training becomes important. Consistently, teachers also displayed favorable attitudes towards sharing among teachers themselves. For example, STA2 noted, "I didn't go for formal course but teachers who have been for the course have come and shared. They have also showed us their worksheets, so it is quite helpful."

Four teachers attributed part of their knowledge of IT to preservice training. Among them, three have taught mathematics for only five years or less. Moreover, three teachers who cited their schooling experience as an important source for them to develop knowledge of IT in the interview are also all from the junior group, STG1.

Similar to the findings on the sources of teachers' knowledge of textbooks, few teachers claimed that their IT knowledge came from their reading of professional journals and books. In the interview, only STB3 mentioned that he borrowed books on the computer from the library to learn Excel.

In addition, some other knowledge sources were suggested by the teachers in the interview, such as previous working experience in IT industry

Table 10.11 Sources of Singaporean teachers' knowledge of technology identified during the interview

	A	B	C	D	E	F	G	Other sources
Frequency of teachers identifying the source	3	4	10	4	6	2	6	6

Note: A = Experience as student, B = Preservice training, C = Inservice training, D = Organized professional activities, E = Informal exchanges with colleagues, F = Reading professional journals and books, G = Own teaching experience and reflection. Other sources: Self-learning (2), Friends (2), Previous working experience (1), and Life experience (1).

(STA1), self-taught through internet (STB3), and outside sales people (STC2 and STF2). It is also interesting to note that STF3 also indicated that she learned some knowledge of using a calculator from her children. Although this appears to be an unusual case, it does tell us that there exist a wide variety of sources for teachers to develop their professional knowledge.

Table 10.11 summarizes the results from the interview about the sources of these Singaporean teachers' knowledge of technology for mathematics teaching.

Knowledge of concrete materials

As mentioned earlier, teachers' knowledge of other teaching resources in the study mainly refers to the knowledge of how to use concrete materials or "physical three-dimensional objects" for teaching mathematics. Available studies on the effectiveness of using concrete materials in mathematics instruction have generally found that use of such concrete materials in mathematics classes was positively related to the improvement of students' mathematics achievement as well as their attitudes towards mathematics.

According to teachers' responses to the questionnaire survey, many more teachers used concrete materials for their mathematics instruction nowadays than their teachers did during their school days (87% vs. 41%). More specifically, the responses to Question 18a revealed that more than 77% of the teachers who responded at least sometimes utilized concrete materials in their mathematics classes, although none of them always

Table 10.12 Results on how Singaporean teachers gave different evaluations on the contribution of various sources to the development of their knowledge of using concrete materials/physical models for teaching mathematics

Sources (Number of teachers)	Contribution				Average evaluation[a]
	Very much	Somewhat	Little	No contribution	
Experience as student (62)	3 (4.84%)	20 (32.26%)	23 (37.10%)	16 (25.81%)	2.16
Preservice training (62)	5 (8.06%)	36 (58.06%)	14 (22.58%)	7 (11.29%)	2.63
Inservice training (61)	5 (8.20%)	33 (54.10%)	17 (27.87%)	6 (9.84%)	2.61
Organized professional activities (61)	5 (8.20%)	26 (42.62%)	24 (39.34%)	6 (9.84%)	2.49
Informal exchanges with colleagues (63)	7 (11.11%)	37 (58.73%)	16 (25.40%)	3 (4.76%)	2.76
Reading professional journals and books (64)	0 (0%)	16 (25%)	29 (45.31%)	19 (29.69%)	1.95
Own teaching experience and reflection (63)	12 (19.05%)	38 (60.32%)	11 (17.46%)	2 (3.17%)	2.95

Link function: Logit.
[a]The average evaluation is obtained by using the ordinal scale, $4 =$ very much, $3 =$ somewhat, $2 =$ little, and $1 =$ no contribution.

utilized concrete materials in mathematics teaching during the school year surveyed.

Table 10.12 shows the detailed results from teachers' responses to Question 18b on teachers' evaluations on the contribution of different sources to the development of their knowledge of how to use concrete materials/physical models for teaching mathematics.

Based on the average evaluation of those teachers on each source shown in Table 10.12, it is clear that the order of importance of the different sources, from the most important to the least important, for teachers to develop their knowledge concerning how to use concrete materials for teaching mathematics is their "own teaching experience and reflection" (2.95), "informal exchanges with colleagues" (2.76), "preservice training" (2.63), "inservice training" (2.61), "organized professional activities" (2.49), "experience as student" (2.16), and "reading professional journals and books" (1.95).

Applying logistic regression analysis reveals the same order of importance of various sources to the development of teachers' knowledge of concrete materials, except for the order of "preservice training" and "inservice training" (note that there is no significant difference between the two sources; see Table 10.13).

Furthermore, Table 10.13 shows that among those sources, the most important one is teachers' "own teaching experience and reflection", the secondarily most important sources are "informal exchanges with colleagues", "preservice training", "inservice training", and "organized professional activities", and the least important ones are teachers' "experience as student" and "reading professional journals and books". Statistically, the three groups of sources contributed significantly differently to teachers' knowledge of concrete materials based on their evaluations.

Applying the chi-square test to the data categorized using the three groups of teachers with different lengths of teaching experience found that

Table 10.13 Results of logistic regression on the data about the contribution of different sources to Singaporean teachers' knowledge of concrete materials

		Estimate	Std. Error	Wald	df	Sig.	95% Confidence Interval	
							Lower Bound	Upper Bound
Threshold	[CONTRIBU = 1]	−.615	.777	.626	1	.429	−2.138	.908
	[CONTRIBU = 2]	.742	.779	.908	1	.341	−.784	2.268
	[CONTRIBU = 3]	3.140	.781	16.179	1	.000	1.610	4.671
Location	[STUDEN = 0]	.987	.196	25.461	1	.000	.604	1.371
	[PRESVC = 0]	.243	.196	1.537	1	.215	−.141	.628
	[ORGACT = 0]	.256	.197	1.695	1	.193	−.130	.642
	[COLLEAG = 0]	−.357	.198	3.267	1	.071	−.745	3.017E-02
	[READING = 0]	1.242	.197	39.601	1	.000	.855	1.629
	[OWNREFL = 0]	−1.204	.203	35.125	1	.000	−1.602	−.805
	[INSVC = 0][a]	0	.	.	0	.	.	.

Link function: Logit.
[a]This parameter is set to zero because it is redundant.

there were no significant differences among the three groups of teachers. In other words, teachers' length of teaching experience did not have significant influences on teachers' evaluation of the importance of those knowledge sources for them to develop their knowledge of using concrete materials to teach mathematics.

During the interview conducted after classroom observations, 15 out of 18 teachers reported that they used concrete materials in mathematics teaching. When asked the sources from which they acquired such knowledge, six teachers attributed part of the knowledge to their teaching experience. However, none of them are from STG1 with five years or less teaching experience. In contrast, three of the six teachers who cited inservice training as one of the knowledge sources are from the junior group. Another three teachers noted that attending organized professional activities were also helpful for them to get to know how to utilize the manipulative models for mathematics teaching.

In addition, no one in the interview mentioned their schooling experiences as a source for them to acquire their knowledge of using concrete materials in mathematics teaching. They also rarely indicated reading professional books and journals as a source.

Table 10.14 provides a summary from the interview about the sources of the Singaporean teachers' knowledge of concrete materials for their teaching of mathematics.

Pedagogical content knowledge

As described earlier, PCnK in this study refers to the knowledge of ways to represent mathematics concepts and procedures, which is an essential part

Table 10.14 Sources of Singaporean teachers' knowledge of concrete materials identified during the interview

	A	B	C	D	E	F	G	Other sources
Frequency of teachers identifying the source	0	2	6	3	3	1	7	1

Note: A = Experience as student, B = Preservice training, C = Inservice training, D = Organized professional activities, E = Informal exchanges with colleagues, F = Reading professional journals and books, G = Own teaching experience and reflection. Other sources: Self-learning.

of mathematics teachers' professional knowledge. As there are infinitely many examples of PCnK in secondary mathematics curriculum and it would be impossible to find common contents that all the participating teachers were teaching, two open-ended questions concerning teachers' PCnK, namely Questions 21 and 22, were particularly designed in the questionnaire. Teachers were first asked to recall two lessons they taught recently that contained new mathematics topics, and then identify the new topics, describe the ways they represented the topics, and finally explain how they got to know the ways they used.

In their responses to the first parts of the two questions, teachers offered a wide spectrum of mathematical topics they had recently taught. In total, there were 108 such topics, which were classified into different mathematical branches based on the Singaporean national mathematics syllabus. They are arithmetic, mensuration, algebra and graphs, geometry, trigonometry, probability and statistics, vectors, matrix and transformation, remainder and factor theorem, differentiation, and simultaneous equations.

Part c of each of the two questions required the teachers to explain how they obtained their knowledge of representing the corresponding topics in the ways they used in their teaching. The overall results, based on our coding of the teachers' descriptions given in the two questions, are shown in Figure 10.2.

From Figure 10.2, we can see that the order of importance of different sources to the development of teachers' PCnK, in terms of the contributions reported by the Singaporean teachers from the most to the least, were their "own teaching experience and reflection" (29.25%), "mathematical resources package (including textbooks)" (22.64%), "inservice training" (16.04%), "experience as student" (13.21%), "preservice training" (13.21%), "informal exchanges with colleagues" (9.43%), "reading professional journals and books" (9.43%), and "organized professional activities" (0.94%).

Logistic regression model was again employed using SPSS to analyze the data in order to detect the significance of different sources to teachers' PCnK. The results are given in Table 10.15.

From Table 10.15, we find that the order of importance of different sources to teachers' developing their PCnK is the same as revealed using the average evaluation, as shown in Figure 10.2.

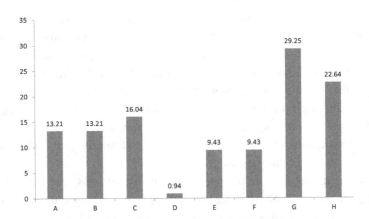

Figure 10.2 Percentage of the Singaporean teachers who got their PCnK about various topics they listed from different sources

Note: 1. A = Experience as student, B = Preservice training, C = Inservice training, D = Organized professional activities, E = Informal exchanges with colleagues, F = Reading professional journals and books, G = Own teaching experience and reflection, H = Mathematics resource package (including textbooks).

2. The total percent of teachers shown in Figure 10.2 is higher than 100% as some teachers provided more than one source of PCnK in their answer.

Table 10.15 Results of logistic regression on the data about the contribution of different sources to the Singaporean teachers' PCnK (binary logistic model)

		B	Std. Error	Wald	df	Sig.	Exp(B)
Step 1	STUDEN	.000	.402	.000	1	1.000	1.000
	INSVC	.223	.386	.332	1	.564	1.249
	ORGACT	−2.753	1.043	6.964	1	.008	.064
	COLLEAG	−.373	.436	.734	1	.392	.689
	READING	−.373	.436	.734	1	.392	.689
	OWNREFL	.968	.352	7.547	1	.006	2.633
	TEXTBOOK	.637	.364	3.058	1	.080	1.891
	Constant	−2.034	.284	51.207	1	.000	.131

Note: Step 1 here indicates that it is the first step (full model) with all predictor variables in it. Variables entered in the model are STUDEN, INSVC, ORGACT, COLLEAG, READING, OWNREFL, and TEXTBOOK. The constant represents the intercept of the model.

Moreover, Table 10.15 also shows that at the 0.05 level, teachers' "own teaching experience and reflection" is significantly more important than "mathematical resources package (including textbooks)", "inservice training", "experience as student", "preservice training", "informal exchanges with colleagues", and "reading professional journals and books", which are in turn significantly more important than "organized professional activities".

It appeared somehow surprising that the results showed that the teachers in Singapore learned the least of their PCnK from "organized professional activities". One possible reason might be that issues on how to teach specific mathematics concepts and procedures did not receive much attention in these kinds of activities in the Singaporean context. The focus might be more on other matters for teachers' professional development. Nevertheless, further investigation is needed in this matter.

Concerning the influence of the length of teaching experience on the contribution of different sources to teachers' PCnK, we again employed chi-square test to the data categorized into the three groups of teachers. The results revealed that there was no significant difference across the three groups about their reporting the contribution of different sources to the growth of their PCnK, except for two sources: "experience as student" and "preservice training".

Tables 10.16 and 10.17 show the numbers of teachers in the three groups giving different reports of the two sources, respectively.

Examining the data given in Tables 10.16 and 10.17 and the corresponding chi-square test results, we can see that at the 0.05 level, significantly

Table 10.16 Singaporean teachers' responses across the three groups to the questions of their developing PCnK from the source of "experience as student"

	Yes	No	Total
STG1	13	40	53
STG2	1	21	22
STG3	0	31	31
Total	14	92	106

Chi-square test: $\chi^2 = 12.083$ df $= 2$ p $= 0.0024$.

Table 10.17 Singaporean teachers' responses across the three groups to the questions of their developing PCnK from the source of "preservice training"

	Yes	No	Total
STG1	12	41	53
STG2	1	21	22
STG3	1	30	31
Total	14	92	106

Chi-square test: $\chi^2 = 8.250 \text{ df} = 2 \text{ p} = 0.0162$.

higher percentages of teachers from STG1 than their colleagues in STG2 and STG3 reported that they acquired PCnK from their "experience as student" and "preservice training". The results suggest that younger teachers have probably benefited from a better education system during their schooling years as well as an improved preservice training package.

In the interview, teachers were further asked about the sources of their PCnK that were identified during the classes observed. All the seven sources listed in the questionnaire survey were mentioned by these teachers interviewed. In particular, 16 out of the 22 teachers believed that they acquired a great deal of PCnK from informal exchanges with their colleagues. In addition, these teachers frequently recognized the contribution from organized professional activities (e.g., seminars, workshops, and department regular meetings) and their reading of professional journals and books.

In contrast, fewer teachers attributed their PCnK to their experiences as students and preservice trainings they received. Nevertheless, taking the length of teaching experience into account, it was found that the teachers who cited preservice experience as a knowledge source were mainly junior teachers.

Table 10.18 presents a summary of the results from the interview about the sources of teachers' PCnK.

Pedagogical instructional knowledge

As mentioned before, teachers' PIK in the study refers to their knowledge about general teaching strategies and classroom organizational techniques.

Table 10.18 Sources of Singaporean teachers' PCnK identified during the interview

	A	B	C	D	E	F	G	Other source
Frequency of teachers identifying the source	6	2	8	76	16	9	28	1

Note: A = Experience as student, B = Preservice training, C = Inservice training, D = Organized professional activities, E = Informal exchanges with colleagues, F = Reading professional journals and books, G = Own teaching experience and reflection. Other source: Textbooks.

Question 6 in the questionnaire was designed to detect the changes in the use of various instructional strategies and classroom organizational models by Singaporean mathematics teachers between the past and nowadays. The data collected from the survey revealed that the most noticeable changes occurred in the use of six strategies, that is, 89% of the teachers participating in the survey reported that they used the strategy of "classroom discussion", however only 38% of them were taught by their previous teachers using the same strategy during their school days, and the corresponding percentages were 80% vs. 31% for "small group work", 85% vs. 7% for "use of computer", 87% vs. 41% for "use of concrete materials", 86% vs. 31% for "hands-on activities", and 82% vs. 32% for "project work".

The above results clearly show that there have been remarkable differences in the way mathematics is taught in Singapore classrooms by the teachers participating in the survey and by those teachers' mathematics teachers in the early days, and many current teachers adopted teaching strategies that they were not exposed to in mathematics classes during their school days. While it appears that the change in some aspects such as "use of computer" could be more due to the technological advancement in the teaching and learning of mathematics, the change in other aspects such as increasingly using "small group work", "hands-on activities", and "project work" also signals, consistently, the change and development of modern pedagogy that teachers have learned from a variety of sources, which is the focus of this study.

Question 20 in the questionnaire was designed to find out how teachers developed their PIK, and the results from the teachers' responses to the question are summarized in Table 10.19.

Table 10.19 Results on how Singaporean teachers gave different evaluations on the contribution of various sources to the development of their PIK

Sources (Number of teachers)	Contribution				Average evaluation*
	Very much	Somewhat	Little	No contribution	
Experience as student (64)	6 (9.37%)	28 (43.75%)	15 (23.44%)	15 (23.44%)	2.39
Preservice training (63)	11 (17.46%)	35 (55.56%)	10 (15.87%)	7 (11.11%)	2.79
Inservice training (64)	12 (18.75%)	40 (62.50%)	10 (15.62%)	2 (3.13%)	2.97
Organized professional activities (62)	8 (12.90%)	32 (51.61%)	15 (24.20%)	7 (11.29%)	2.66
Informal exchanges with colleagues (65)	13 (20%)	40 (61.54%)	11 (16.92%)	1 (1.54%)	3.00
Reading professional journals and books (65)	2 (3.08%)	24 (36.92%)	28 (43.08%)	11 (16.92%)	2.26
Own teaching experience and reflection (65)	26 (40%)	36 (55.38%)	3 (4.62%)	0 (0%)	3.35

*The average evaluation is obtained by using the ordinal scale, $4 = $ very much, $3 = $ somewhat, $2 = $ little, and $1 = $ no contribution.

The results in Table 10.19 show that there are a variety of sources for teachers to develop their PIK. Using teachers' average evaluation on the contribution of each source, we can see that the order of importance of different sources to the development of teachers' PIK, from the most important to the least important, were their "own teaching experience and reflection" (3.35), "informal exchanges with colleagues" (3.00), "inservice training" (2.97), "preservice training" (2.79), "organized professional activities" (2.66), "experience as student" (2.39), and "reading professional journals and books" (2.26).

Applying logistic regression analysis to the data suggests that the order of importance of the different sources to the development of teachers' PIK is the same as that revealed using the average evaluation shown in Table 10.19. Furthermore, at the 0.05 level, teachers' "own teaching experience and

Table 10.20 Results of logistic regression on the data about the contribution of different sources to the Singaporean teachers' PIK

		Estimate	Std. Error	Wald	df	Sig.	95% Confidence Interval Lower Bound	95% Confidence Interval Upper Bound
Threshold	[CONTRIBU = 1]	−2.102	.668	9.901	1	.002	−3.411	−.793
	[CONTRIBU = 2]	−.731	.668	1.197	1	.274	−2.039	.578
	[CONTRIBU = 3]	1.736	.666	6.790	1	.009	.430	3.041
Location	[STUDEN = 0]	.785	.167	22.003	1	.000	.457	1.113
	[INSVC = 0]	−.283	.170	2.760	1	.097	−.617	5.091E-02
	[ORGACT = 0]	9.667E-02	.169	.326	1	.568	−.235	.429
	[COLLEAG = 0]	−.557	.171	10.645	1	.001	−.891	−.222
	[READING = 0]	1.044	.168	38.548	1	.000	.715	1.374
	[OWNREFL = 0]	−1.435	.176	66.793	1	.000	−1.779	−1.091
	[PRESVC = 0][a]	0	.	.	0	.	.	.

Link function: Logit.
[a]This parameter is set to zero because it is redundant.

reflection" and "informal exchanges with colleagues" are significantly more important than the sources of "inservice training", "preservice training", and "organized professional activities", which are in turn significantly more important than the sources of "experience as student" and "reading professional journals and books" (see Table 10.20).

Applying the chi-square test to the data reveals that the length of the teachers' teaching experience did not affect their viewing of the contributions of the different sources to the development of their PIK. In other words, the length of teaching experience does not have significant influence on teachers' view about the importance of the various sources in the development of their PIK.

Like the data from the questionnaire survey, the data gathered from classroom observation also suggested that teachers employed a variety of general instructional strategies and classroom organizational models in their teaching. Concerning the teaching strategies and classroom management techniques that teachers used in the classroom observations, we later asked them relevant questions about the sources from which the teachers got to know the relevant skills in the interview.

Consistent with the findings from the questionnaire survey, teachers' "own teaching experience and reflection" was cited as a knowledge source most frequently by the teachers. In particular, all but three teachers explained that they learnt a great deal of PIK from their daily teaching. Teachers also indicated that informal exchanges with colleagues, attending organized professional activities, and formal training were very helpful for them to acquire PIK, particularly on new skills. Although some teachers believed they could also get useful ideas from professional reading, due to heavy workload and limited time, teachers seldom resorted to this source to upgrade their PIK. In this aspect, HODs could play an important role in collecting and distributing related articles from professional journals and books for teachers. In fact, in the interview, four out of the six HODs claimed that they did so.

A summary of the results from the interview is given in Table 10.21.

As it is often said that teachers taught the way they were taught, we intentionally asked all the teachers what they thought about this general belief. As a result, five teachers completely agreed with this view, while six teachers totally disagreed with it. The other teachers partially agreed with this opinion and some of them stressed that the ways their teachers adopted to teach them had important influence on them during the early years of their teaching career. However, over the years, as many things including mathematics contents and students are changing, it is necessary for teachers to modify, if not totally give up, the old ways that are familiar to them and employ new ways that are unfamiliar to them for teaching mathematics.

Incidentally, most teachers interviewed in this study expressed that they have changed their teaching strategies in the recent years. They

Table 10.21 Sources of Singaporean teachers' PIK identified during the interview

	A	B	C	D	E	F	G	Other sources
Frequency of teachers identifying the source	9	13	11	9	8	4	23	2

Note: A = Experience as student, B = Preservice training, C = Inservice training, D = Organized professional activities, E = Informal exchanges with colleagues, F = Reading professional journals and books, G = Own teaching experience and reflection. Other sources: MOE Service (1) and Other school (1).

had incorporated more IT and applied different problem solving strategies in their teaching. They confirmed that the major sources for them to make changes are their own teaching experience and reflection, informal exchanges with colleagues, and inservice training provided by the schools and the MOE.

Summary of the Singapore Study

Like the Chicago study, the Singapore study is also centered on teachers' knowledge development, which is a crucial component of teachers' professional development in modern education. More specifically, the study looked into how Singaporean school mathematics teachers developed their knowledge in the domain of pedagogy, which is a key component of teachers' professional knowledge.

A total of 73 mathematics teachers in six secondary schools, a stratified random sample from all of the 152 secondary schools in Singapore, participated in this study. Three instruments were designed and used to collect the data from the research sample, including a questionnaire survey, classroom observations, and follow-up interviews. Both quantitative and qualitative methods of analysis were employed to analyze the data.

Table 10.22 provides a summary of the main findings of this study. The results about the significant differences in the importance of different sources for teachers to develop various components of their pedagogical knowledge were obtained using logistic regression analysis, and the significant differences in the importance of the same source to different groups of teachers were obtained using the chi-square test.

Like the Chicago study, the Singapore study also reveals that there are various sources from which teachers can develop their pedagogical knowledge. Overall, teachers' "own teaching experience and reflection" and "informal exchanges with colleagues" are the most important source for teachers to develop their pedagogical knowledge. Teachers' "inservice training", "preservice training", and "organized professional activities" are secondarily important sources. In contrast, teachers' "experience as student" and their "reading professional journals and books" are the least important ones. One exception is for teachers' PCnK, for which teachers gained more from "experience as student" and "reading professional

Table 10.22 A summary of the main findings on the relative importance of different sources to the development of Singaporean teachers' pedagogical knowledge

Sources[a]	PCrK of			PCnK[b]	PIK
	Teaching materials (textbooks)	Technology	Concrete materials		
Most important	G, E	G, E	G	G	G, E
Secondarily important	C, B, D, A, F	C, D, B	E, C, B, D	C, A,[c] B,[c] E, F	C, B, D
Least important		A, F	A, F	D	A, F

[a] A = Experience as a student, B = Preservice training, C = Inservice training, D = Organized professional activities, E = Informal exchanges with colleagues, F = Reading professional journals and books, G = Own teaching experience and reflection.
[b] Textbooks are also a secondarily important source.
[c] These sources are significantly more important to younger teachers than to senior teachers.

journals and books" than from "organized professional activities". In addition, mathematics textbooks are also a very important source from which teachers can develop their PCnK.

The Singapore study also suggests that, in general, the length of teaching experience does not have a significant influence on teachers' views about the importance of various sources in developing their pedagogical knowledge. Nevertheless, it appears that younger teachers benefited more from their experiences as school students and preservice training than their senior colleagues.

The results of the Singapore study also have several important implications concerning teacher professional development in Singapore.

First, it seems necessary for teacher education providers to further reform preservice education programs and equip prospective teachers with a greater range of pedagogical knowledge and newer skills.

Second, more attention should be paid to the importance of "inservice training" in developing teachers' pedagogical knowledge. In particular, necessary support from various aspects (e.g., educational policy makers, teacher educators, and school administrators) should be given for teachers to pursue inservice training to upgrade their knowledge and skills so that they could conform to the fast changing working environment in their teaching career.

Third, it would be very desirable for school principals and HODs to promote teachers' daily informal exchanges with their colleagues in seeking teachers' professional development. As the study shows it is a most important source for teachers to gain their pedagogical knowledge, and teachers learnt much from such daily exchanging.

Fourth, teachers need to be lifelong learners of how to teach more effectively, as their own experiences and reflection play the most important role in their professional development. More specifically, teachers should always reflect on their own teaching practices, intentionally establish a store of pedagogical knowledge, regularly exchange ideas with colleagues, and actively seek all available sources to develop professionally.

Finally, it should be pointed out that the main findings obtained from the Chicago study and the Singapore study are largely consistent. In particular, in both studies, teachers' own teaching experience and reflection was found to be the most important source for all teachers to develop all the components of their pedagogical knowledge, which appears to a large degree to reveal the nature and requirement of teaching as a profession and teachers' professional development. However, there also exist some differences. In particular, for the development of all the components of teachers' pedagogical knowledge, preservice training was found to be a secondarily important source to mathematics teachers in Singapore but a least important source to their counterparts in Chicago; it seems that the reason is related to the fact that the US and Singapore have very different preservice training systems and practices for teachers. More detailed comparison and analysis are offered in the next chapter.

Chapter 11

Comparison and Conclusion

When the Singapore study was designed, one of its main purposes was to make a comparison with the Chicago study and find out the similarities and differences between the two places with different social, cultural, and educational contexts in relation to the development of teachers' knowledge of pedagogy. This chapter is devoted to this purpose, taking a comparative look at the similarities and differences as revealed in the two studies. Some concluding remarks are offered at the end of the chapter.

Comparison of the Chicago and Singapore Studies

In this section, we first compare the results of the two studies below in terms of the three components of teachers' pedagogical knowledge, that is, pedagogical curricular knowledge (PCrK), pedagogical content knowledge (PCnK), and pedagogical instructional knowledge (PIK), respectively. Finally, we compare the results about the differences among the teachers of three groups in terms of the length of their teaching experiences found in the two studies.

Teachers' pedagogical curricular knowledge

As described earlier, teachers' PCrK in the studies refers to their knowledge of instructional materials and teaching resources. It is investigated through its three sub-components, i.e., knowledge of teaching materials (primarily textbooks), knowledge of technology (primarily calculator and computer/software), and knowledge of other teaching resources (primarily concrete materials).

Table 11.1 Comparison of the contribution of different sources to the development of teachers' curricular knowledge in the Chicago and Singapore studies

	PCrK					
	of textbooks		of technology		of concrete materials	
Sources	Chicago	Singapore	Chicago	Singapore	Chicago	Singapore
Most important	G, E	G, E	G, E, C	G, E	G, E	G
Secondarily important	C, D, A	C, B, A, D, F	D	C, D, B, A	C, D	E, B, C, D
Least important	F, B		F, A, B	F	F, B, A	A, F

Note: For the sources, A = Experience as student, B = Preservice training, C = Inservice training, D = Organized professional activities, E = Informal exchanges with colleagues, F = Reading professional journals and books, G = Own teaching experience and reflection.

Table 11.1 provides a comparison of teachers' evaluations of the importance of the different sources to the development of their curricular knowledge in the Chicago and Singapore studies, based on the relevant results of logistic regression analysis and the chi-square test as reported in Chapter 5 and Chapter 10 respectively. The order of the seven sources is the same as that based on the teachers' average evaluation for each source, from high to low, by using the ordinal scale, that is, 4 = very much, 3 = somewhat, 2 = little, and 1 = no contribution, as stated earlier.

From the results as shown in Table 11.1, we can see the main similarities as follows:

First, Source G — "own teaching experience and reflection" — and Source E — "informal exchanges with colleagues" — are the two most important sources for teachers' development of their PCrK in all three components in both places, except in Singapore Source E is a secondarily important source in the component knowledge of concrete materials (though it is still the second most important source in terms of teachers' average evaluation). The consistent results revealed in both places not only explain the importance of teachers' experience and reflection, and their daily exchange with colleagues to the development of their pedagogical knowledge, but also reveal the nature of teaching as a profession.

Second, Source C — "inservice training" — and Source D — "organized professional activities" — played a secondarily important role in teachers' development of their PCrK in all the three components in both places, except in the component knowledge of technology Source C is a most important source in the case of Chicago. The similarities suggest the value of external organized help for teachers' need in developing their knowledge.

Third, Source F — "reading professional journals and books" — was regarded as the least important source for teachers' PCrK in all the three components in both places, except in the component of knowledge of technology in Singapore, in which it is a secondarily important source. Nevertheless, in terms of teachers' average evaluation, it is still the least important source. The result is not surprising, as teachers in both places normally need to teach a big number (often 30 to 40 hours) of lessons besides other official duties, and therefore they have difficulty finding time to read professional journals and books.

The main differences exist among Sources A and B. While Source A, "experience as student", is a secondarily important source in developing teachers' knowledge of textbooks and a least important source in developing their knowledge of concrete materials in both places, it is a least important source in Chicago but a secondarily important source in Singapore in developing teachers' knowledge of technology.

The biggest difference is found in Source B, that is, "preservice training". In Chicago, it was viewed as a least important source for all the three components, in Singapore it was regarded as a secondarily important source for all the components. To me, this largely points to the possibility that the preservice training in the US is less effective than that in Singapore for preservice teachers gaining PCrK, though to make a detailed comparison about the effectiveness of preservice training in these two countries is beyond the scope of the two studies.

Table 11.2 summarizes the value of teachers' average evaluation of contribution of each source to the development of their knowledge in the three components.

The figures in Table 11.2 reveal that Source A, "experience as student", and Source B, "preservice training", received consistently higher average evaluation by teachers in Singapore across the three components than their

Table 11.2 Comparison of teachers' average evaluation of the contribution of different sources to teachers' PCrK in the two studies

PCrK	Study	Source						
		A	B	C	D	E	F	G
of textbooks	Chicago	1.66	1.46	2.15	1.88	3.32	1.50	3.81
	Singapore	2.05	2.11	2.16	2.07	2.59	1.91	3.18
of technology	Chicago	1.72	1.66	3.36	3.03	3.63	2.10	3.72
	Singapore	2.14	2.59	2.92	2.80	2.91	2.02	3.25
of concrete materials	Chicago	1.64	1.87	2.54	2.49	3.33	2.01	3.33
	Singapore	2.16	2.63	2.61	2.49	2.76	1.95	2.95

Note: 1. For the sources, A = Experience as student, B = Preservice training, C = Inservice training, D = Organized professional activities, E = Informal exchanges with colleagues, F = Reading professional journals and books, G = Own teaching experience and reflection.
   2. The evaluations are shown by the ordinal scale in the figure, 4 = very much, 3 = somewhat, 2 = little, and 1 = no contribution.

counterparts in Chicago; Source E, "informal exchange with colleagues" and Source G, "own experience and reflection" received consistently lower average evaluation in Singapore than in Chicago; while Source C, "inservice training", Source D, "organized professional activities", and Source F, "reading professional journals and books", received a mixed average evaluation.

Teachers' pedagogical content knowledge

With regard to how teachers developed their PCnK, Table 11.3 displays the relative frequency (%) of teachers identifying different sources for their PCnK about the teaching of specific mathematics contents in Chicago and Singapore, as reported earlier. The differences of the relative frequencies between the two places for each source identified are added in the table.

Considering the relative frequency shown in Table 11.3, we can see that there are obvious differences in terms of how teachers viewed where they gained their pedgogical content knowledge from different sources in Chicago and Singapore. Sources A, "experience as student", B, "preservice training", C, "inservice training", F, "reading professional journals and books", and H, "textbooks", were all more highly evaluated by teachers in

Table 11.3 Comparison of the relative frequency (%) of teachers identifying different sources for their PCnK about the teaching of specific mathematics contents

	Source							
	A	B	C	D	E	F	G	H
Chicago	5.6%	2.8%	5.6%	11.1%	33.3%	5.6%	50.0%	19.4%
Singapore	13.2%	13.2%	16.0%	0.9%	9.4%	9.4%	29.3%	22.6%
Differences	−7.6%	−10.4%	−10.4%	10.2%	23.9%	−3.8%	20.7%	−3.2%

Note: 1. For the sources, A = Experience as student, B = Preservice training, C = Inservice training, D = Organized professional activities, E = Informal exchanges with colleagues, F = Reading professional journals and books, G = Own teaching experience and reflection, H = Textbooks.

2. The total percentage of teachers shown in the table is higher than 100% as some teachers provided more than one source of their PCnK in their answer.

Table 11.4 Comparison of teachers' evaluations of the contribution of different sources to teachers' PCnK

	Most important	Secondarily important	Least important
Chicago	G, E	H, D, A, C, F	B
Singapore	G	H, C, A, B, E, F	D

Note: For sources, A = Experience as student, B = Preservice training, C = Inservice training, D = Organized professional activities, E = Informal exchanges with colleagues, F = Reading professional journals and books, G = Own teaching experience and reflection, H = Textbooks.

Singapore in relation to the development of their PCnK, while Sources G, "own teaching experience and reflection", E, "informal exchanges with colleagues", and D, "organized professional activities", were viewed to be more imprtant in Chicago than in Singapore.

Furthermore, we also notice that the biggest differences of the teachers' views in the two places were found in relation to Sources G, E, B, C, and D, all more than 10%.

Table 11.4 summarizes the relevant importance levels of the different sources obtained in the two studies, which is based on the relevant results of logistic regression analysis and the chi-square test as reported in Chapter 6 and Chapter 10, respectively.

From Table 11.4, we can clearly see that the main similarities are found in Sources G, H, A, C, and F. In both Chicago and Singapore, Source G, "own teaching experience and reflection", was considered the most important source for teachers' development of their PCnK, while Source H, "textbooks", Source A, "experience as student", Source C, "inservice training", and Source F, "reading professional journals and books", were viewed as the secondarily important sources.

The main differences exist among Sources E, D, and B. While Source E, "informal exchanges with colleagues", was found in the Chicago study to be one of the most important sources, ranked second in terms of teachers' average evaluation, it was found in the Singapore study to be only a secondarily important source, ranked sixth.

Source D, "organized professional activities", was a secondarily important source, ranked third in Chicago, but in Singapore it was viewed as a least important source, ranked seventh, or the very last. From the result, it seems reasonable to say that organized professional activities in the Chicago schools were more helpful in developing teachers' PCnK, while in Singapore schools this kind of knowledge sharing and development probably did not receive much attention in organized professional activities, an issue worth further investigation and action for Singapore schools and educators.

Source B, "preservice training", again was considered to be a more important source in Singapore than in Chicago for teachers' gaining of their PCnK, which I think is another indication of the effectiveness of the preservice training in both places in preparing teachers' professional knowledge, an issue calling for more attention and action from the US mathematics teacher educators.

Teachers' pedagogical instructional knowledge

PIK is about general teaching strategies and classroom organizational models, as explained earlier in the book. For the purpose of comparison, we again first look at the relevant importance levels of the different sources based on the relevant results of logistic regression analysis as reported in Chapter 7 and Chapter 10 respectively, as summarized in Table 11.5. Again, it should be emphasized that all the data were collected from teachers' evaluations in the questionnaire surveys.

Table 11.5 Comparison of the contribution of different sources to teachers' PIK in Chicago and Singapore

	Most important	Secondarily important	Least important
Chicago	G, E, C	D	B, A, F
Singapore	G, E	C, B, D, A	F

Note: For the sources, A = Experience as student, B = Preservice training, C = Inservice training, D = Organized professional activities, E = Informal exchanges with colleagues, F = Reading professional journals and books, G = Own teaching experience and reflection, H = Textbooks.

From Table 11.5, it is clear that the similarities about the importance levels of different sources between the Chicago and Singapore studies exist in Sources G, E, D, and F. More specifically, Source G, "own teaching experience and reflection", and Source E, "informal exchanges with colleagues", were consistently viewed by the teachers in the two places to be the most important sources for their PIK, Source D, "organized professional activities", was a secondarily importance source, and Source F, "reading professional journals and books", a least important source.

The main differences are found in Sources C, B, and A. According to teachers' average evaluation, while Source B, "preservice training", and Source A, "experience as student", are secondarily important sources in Singapore, they have least importance in Chicago, and moreover, Source C, "inservice training", is a most important source in Chicago but a secondarily important one in Singapore.

Table 11.6 shows teachers' average evaluation about the contribution of different sources to the development of their PIK. Differences are also added in the table.

The descriptive statistics shown in Table 11.6 reveal that Source E, "informal exchanges with colleagues", and Source G, "own teaching experience and reflection", were more favorably evaluated by teachers in Chicago than those in Singapore, even though both sources were regarded as the most important sources compared to other sources, as shown in the logistic model analysis. In contrast, Source B, "preservice training", was more favorably evaluated by teachers in Singapore than their counterparts in Chicago.

Table 11.6 Comparison of teachers' average evaluation of the contribution of different sources to teachers' PIK in Chicago and Singapore

	Source						
	A	B	C	D	E	F	G
Chicago	2.20	2.28	2.97	2.67	3.70	2.10	3.86
Singapore	2.39	2.79	2.97	2.66	3.00	2.26	3.35
Differences	−0.19	−0.51	0	0.01	0.70	−0.16	0.51

Note: 1. For the sources, A = Experience as student, B = Preservice training, C = Inservice training, D = Organized professional activities, E = Informal exchanges with colleagues, F = Reading professional journals and books, G = Own teaching experience and reflection.

2. The evaluations are shown by the ordinal scale in the figure, 4 = very much, 3 = somewhat, 2 = little, and 1 = no contribution.

In addition, there are no or slight differences regarding other sources, that is, Source C, "inservice training", Source D, "organized professional activities", Source F, "reading professional journals and books", and Source A, "experience as student".

Comparison of three groups of teachers

As described earlier, both Chicago and Singapore studies were also designed in part to detect the possible differences in the importance of different sources to the development of pedagogical knowledge among teachers with different lengths of teaching experience. For this purpose, teachers were classified into three groups with teaching experiences of 0–5 years, 6–15 years, and 16 or more years, respectively.

Table 11.7 summarizes the results about the differences among the teachers of three groups found in the two studies.

Examining the table, we can see that, for PCrK, the only differences were found in the Chicago study in that Sources A, "experience as student", and Source B, "preservice training", are significantly more important to younger teachers than to senior teachers in developing teachers' knowledge of technology, and Source B is also significantly more important to younger teachers in developing their knowledge of concrete materials. Also in the Chicago study, Source E, "informal exchange with colleagues", was

Comparison and Conclusion 269

Table 11.7 Differences about the importance of each source to the development of pedagogical knowledge among the teachers of three groups in Chicago (C) and Singapore (S)

		A	B	C	D	E	F	G
PCrK	of textbook	—	—	—	—	—	—	—
	of technology	C: >	C: >	—	—	—	—	—
	of concrete materials	—	C: >	—	—	—	—	—
PCnK		S: >	S: >	—	—	C: >	—	—
PIK		—	—	—	—	—	—	—

Note: 1. For the sources, A = Experience as student, B = Preservice training, C = Inservice training, D = Organized professional activities, E = Informal exchanges with colleagues, F = Reading professional journals and books, G = Own teaching experience and reflection.

2. ">" means that the corresponding source is significantly more important to younger teachers than to senior teachers, and "—" means there is no significant difference.

significantly more important to younger teachers than to senior teachers in developing their PCnK. But all these differences were not found in the Singapore study. Here, Source A, "experience as student", and Source B, "preservice training", have a more significant importance to younger teachers than to senior teachers in this aspect in developing their PCnK.

Overall, it appears clear from Table 11.7 that the length of teachers' teaching experiences largely bears no significant importance on the different sources on teachers' gaining pedagogical knowledge. The results also suggest that, more often than not, teaching is a regularly changing profession consistently demanding new knowledge, and teachers need to be lifelong learners.

Concluding Remarks

Table 11.8 provides an overview of the results of the Chicago and Singapore studies based on the statistical analysis.

A few concluding remarks are in order to end this book.

The first is about Source G, "Own teaching experience and reflection", and Source E, "Informal exchanges with colleagues". When I planned the Chicago study, I did not expect that teachers' own experience and reflection

Table 11.8 An overview of the results of the Chicago (CHA) and Singapore (SING) studies

| | PCrK | | | | | | PCnK[b] | | PIK | |
| | of textbooks | | of technology | | of concrete materials | | | | | |
Sources[a]	CHA	SING	CHA	SING	CHA	SING	CHA	SING	CHA	SING
Most important	G, E	G, E	G, E, C	G, E	G, E	G	G, E[c]	G	G, E, C	G, E
Secondarily important	C, D, A	C, B, A, D, F	D	C, D, B, A	C, D	E, B, C, D	D, A, C, F	C, A,[c] B,[c] E, F	D	C, B, D, A
Least important	F, B		F, A,[c] B[c]	F	F, B,[c] A	A, F	B	D	B, A, F	F

[a] A = Experience as student(s), B = Preservice training, C = Inservice training, D = Organized professional activities, E = Informal exchanges with colleagues, F = Reading professional journals and books, G = Own teaching experience and reflection.
[b] "Textbooks" are also a secondarily important source in both Chicago and Singapore.
[c] These sources are significantly more important to younger teachers than to senior teachers.

and their daily or informal exchange with colleagues played such an important role in teachers' gaining their pedagogical knowledge. After I completed the study, from the statistical analysis at the macro-level as well as from my interviews with teachers and listening to their in-depth delineations at the micro-level about how and what they learned from different sources, this result was not surprising at all to me as a researcher.

Following the Chicago study, when I planned the Singapore study, the above result was to a certain degree like a hypothesis. The hypothesis was confirmed in the Singapore study for all groups of teachers and for all components of pedagogical knowledge except that teachers' daily exchange with colleagues in Singapore was a secondarily important source for teachers to gain their PCnK, which appears to suggest that compared with Chicago, in Singapore teachers' daily exchanges with their colleagues focused to some degree less on how to teach specific mathematics contents.

Both studies show clearly that teachers' own teaching experience and reflection is the most important source for their knowledge of pedagogy. It seems to me that this consistent result reveals not only the nature of teaching as a profession, that is, teachers must become a reflective learner and learn on the job and from their experience, but also the nature of the knowledge of pedagogy, that is, much of the pedagogical knowledge that teachers need is tacit, practice-oriented, and classroom-based about the real students, curriculum, and classroom, which is regularly if not constantly changing, implying that teachers also need to be lifelong learners.

As mentioned in Chapter 9, for educational policy makers and school administrators, the result also indicated the importance of teacher retention in improving the quality of teaching and learning in schools, as it is unrealistic to expect new teachers to have substantial teaching experiences before they join the profession. In a large sense, the issue of teacher attrition is an issue of loss of knowledge in teachers.

Second, it is about preservice training. When I planned the Chicago study, I thought that preservice training would be a most important source. After all, no one would deny that a main purpose of preservice training, or university teacher education programs, should be to equip preservice trainee teachers with adequate professional knowledge for their teaching career.

The result from the Chicago study is most disappointing to me as a teacher educator as it was considered by the teachers to be the least important source for them to gain their pedagogical knowledge in all components. In contrast, in Singapore preservice training was viewed significantly more important, though it is still a secondarily important source, in all the components of teachers' pedagogical knowledge.

The results from both the Chicago and Singapore studies suggest a repositioning of preservice training in teacher education and professional development is necessary. To teacher educators working in higher educational institutions, there is a need to extend preservice training to inservice training, as the value of preservice training should not be overestimated. To teachers, they must realize that getting a teaching qualification or license from preservice training programs is only a starting point for their pedagogical knowledge, and it is far from sufficient, let alone enough. And to educational policy makers and school administrators, effective measures must be taken to support new teachers' continuous learning on the job.

Comparing the Chicago and Singapore studies, I think the results also present a clear and bigger challenge for the US policy makers and teacher educators to reform and improve its teacher education system and practice. Since Singapore students' stellar performance in TIMSS studies became well known, many US researchers have paid much attention to Singapore's education system. The results from the studies reported in this book indicate that the effectiveness of preservice training in the US is considerably lower compared with that in Singapore, an issue worth further attention and investigation.

Thirdly, it is about inservice training. Compared with preservice training, both Chicago and Singapore studies revealed that it is either a significantly more important or an equally important source for the development of teachers' pedagogical knowledge. Particularly in Chicago, inservice training was considered to be one of the most important sources in teachers' knowledge of technology and their PIK. The results clearly indicate the importance of inservice training for teachers' knowledge development. As this author has pointed out elsewhere, the reasons appear to be related to inservice training's unique advantages including inservice teachers' better professional background, relatively strong practical need and motivation, and the immediate applicability of the knowledge they learn in the training to their actual classroom teaching (Fan, 2002).

Fourthly, it is about organized professional activities. In Chicago, it was a secondarily important source and was also consistently significantly more important than preservice training, while in Singapore, it was a secondarily important source and an equally important source for all the components of teachers' pedagogical knowledge, except it was the least important source and significantly less important than preservice training for teachers to develop their PCnK. In a sense, the result suggests that Singapore schools need to organize professional activities more effectively, and in particular, pay more attention to the issues about how to teach specific mathematics contents.

Taking into account both inservice training and organized professional activities and comparing the findings in Chicago and Singapore, it is clear that educational policy makers, school administrators, and teacher educators should pay reasonable attention to inservice training and organized professional activities so they can play an important role in the development of teachers pedagogical knowledge.

Fifthly, it should also be noted that compared with Chicagoan teachers, Singaporean teachers' average evaluation of the importance of their own teaching experience and reflection and their daily exchange with colleagues were consistently lower to the development of their pedagogical knowledge, even though in both places, these two sources were considered to be overall the most important sources.

It will also be interesting to further investigate if such difference is due to the less effectiveness as reported in the Chicago study about preservice training the teachers received so they need to resort more to these two sources, or even inservice training and organized professional activities, as mentioned above, on the job in order to compensate for the inadequate preparation in their preservice training, while in Singapore the opposite is the case, or whether it is more due to the different school cultures and teachers' professional development practices in the two places. However, no matter what the reasons are, it seems clear that more Singaporean teachers need to be aware of the importance of learning on the job, from reflection, and from exchange with colleagues.

Sixthly, both the studies revealed that reading professional journals and books was a least important source for teachers to develop their pedagogical knowledge. Considering that teachers in both Chicago and Singapore usually need to teach on average about 30–35 class periods per week in addition to many other official tasks, the result is not really surprising. It indicates the dilemma between the need for teachers' knowledge development and the limited amount of time available due to their busy teaching load. I think this is an issue worth further attention from the educational policy makers and school administrators.

Incidentally, the result that experience as students was overall viewed as one of the least important sources in teachers' pedagogical knowledge suggest that the widely held belief that teachers teach the way they were taught was largely untrue, as mentioned earlier.

Finally, as many researchers have noted, there has been much debate over the last decades about the nature of teaching as a profession, professional development, and professional practice (e.g., see Edwards & Nicoll, 2006), and the issue as to how teachers develop their professional knowledge, simply termed by some researchers "teacher learning", continues to be a hot issue (e.g., see Cooney, 1999; Zeichner, 1999; Darling-Hammond & Bransford, 2005; Sztein, 2010). It is also worth mentioning that the theme of the American Educational Research Association 2004 Annual Meeting was "teacher learning and development", which was claimed to be a critically important and timely issue in educational research and discourse from a broad perspective (Borko & Putnam, 2003), and in the area of mathematics education, The International Association for Evaluation of Educational Achievement (IEA) has expanded its focus of comparison from students to teachers, and conducted a large-scale comparative study on mathematics teacher education, The Teacher Education and Development Study in Mathematics or TEDS-M, which included issues about teachers' knowledge of mathematics and teaching of mathematics (e.g., see Tatto *et al.*, 2008), although the study was mainly limited to prospective mathematics teachers and preservice teacher education. More recently, while Sadler *et al.* (2013) provided new evidence about the importance of teachers' knowledge for higher student gain in science learning, Boyd *et al.* (2013) argued "what is less clear is how teachers develop such [mathematical] knowledge for teaching". It is the hope of this author that the two studies presented in this book can contribute meaningfully to the understanding and on-going discourse concerning how teachers develop their professional knowledge. Obviously, given the scope and limitation of the studies, there is much more to be further explored in this important issue of education.

Appendix 1A

Teacher Questionnaire (Chicago Study)

Name (optional)_____ School_____ Date_____

The Sources of Teachers' Pedagogical Knowledge

> This is part of a study into how teachers developed their knowledge of how to teach mathematics. The purpose of the study is to provide guidance to schools and teacher-training institutions. All responses will be kept strictly confidential; neither schools nor teachers will be identified.
>
> Your response is very important to us. Please respond to all questions as best as you can. As a way to express our thanks, we shall send a brief report of the study, when available, to each participant if he/she wishes. Please check "Yes" or "No" below.
>
> **Do you wish to receive a brief report of the study?** ☐ Yes ☐ No
>
> We sincerely appreciate your cooperation.

A. Your background information.

1. Gender: ☐ Male ☐ Female
2. Age: ☐ <20 ☐ [20, 30) ☐ [30, 40) ☐ [40, 50) ☐ [50, 60)
   ☐ ≥60

3. Degrees earned:

Degree	Major	Minor(s)
Associate's	_____	_____
Bachelor's	_____	_____
Master's	_____	_____
Doctorate	_____	_____

4. For how many years have you taught: Any subjects ____years
   Mathematics* ____years

   including statistics, but excluding computer courses

> **B.** Recall your experience as a student in k-12 math classes and your experience of teaching.

5. Which of the following strategies did you encounter as a student, or use as a teacher?

Teaching strategies	As student being taught (used by your teachers)			As teacher teaching (used by yourself)	
	Yes	No	DR (Don't remember)	Yes	No
a. review of previous lessons	Yes	No	DR	Yes	No
b. lecture on new topics	Yes	No	DR	Yes	No
c. classroom discussion	Yes	No	DR	Yes	No
d. small group work	Yes	No	DR	Yes	No
e. reading assignment from the texts	Yes	No	DR	Yes	No
f. use of calculator	Yes	No	DR	Yes	No
g. use of computer	Yes	No	DR	Yes	No
h. use of concrete materials	Yes	No	DR	Yes	No
i. hands-on activities	Yes	No	DR	Yes	No

> **C.** By "pre-service training" we mean teacher-preparation training (e.g., a teacher education program) you received <u>before the first time</u> you became a school teacher.

6. Did you receive pre-service training? ☐ Yes ☐ No
   If "No", please skip to Question 10.
   If "Yes", the training was for   ☐ teaching mathematics
   ☐ teaching other subjects.

7. During your pre-service training, were you taught the following knowledge or skills?

a. how to lecture	Yes	No	DR
			(Don't Remember)
b. how to manage classrooms	Yes	No	DR
c. how to use computers for teaching math	Yes	No	DR
d. how to use calculators for teaching math	Yes	No	DR
e. how to use concrete materials for teaching math	Yes	No	DR
f. *programmed* learning	Yes	No	DR
g. *discovery* learning	Yes	No	DR
h. *cooperative* learning	Yes	No	DR
i. *constructivism* in math learning	Yes	No	DR
j. how to teach specific math topics	Yes	No	DR

8. How useful were the following courses of your pre-service training in enhancing your knowledge of *how to teach mathematics*?

	Very useful	Useful	Not very useful	Not useful	DNT (Did not take)
General educational courses	4	3	2	1	DNT
Mathematics methods courses	4	3	2	1	DNT
Student teaching practice	4	3	2	1	DNT

9. Overall, how useful was your pre-service training in enhancing your knowledge of *how to teach mathematics?*
☐ Very useful   ☐ Useful   ☐ Not very useful   ☐ Not useful

D. Now we turn to professional training you received <u>after the first time</u> you became a teacher. "Professional training" is *different from* other general "professional activities", such as <u>general</u> conferences and classroom observation, which do NOT have specific *"training"* purposes.

---

10. Since becoming a teacher, have you taken college/university courses for credit which were <u>not</u> for a degree?   ☐ Yes   ☐ No
    If yes, those courses are in the field(s) of _____

    Why did you take them? (check all that apply)
    ☐ Required by the State   ☐ Required by the district
    ☐ My own choice   ☐ Other_____

    How useful were they in enhancing your knowledge of *how to teach mathematics?*
    ☐ Very useful   ☐ Useful   ☐ Not very useful   ☐ Not useful

11. Since becoming a teacher, have you enrolled in one or more degree programs?
    ☐ Yes   ☐ No
    If yes, please list the degree(s) you earned or you are earning. Then evaluate the usefulness of the degree program(s) in enhancing your knowledge of *how to teach mathematics?*
    (4 = Very useful   3 = Useful   2 = Not very useful   1 = Not useful)

Degree & *Major*	**Usefulness**
_____	4 3 2 1
_____	4 3 2 1

    Why did you enroll in the degree program(s)? (check all that apply)
    ☐ Required by the State   ☐ Required by the district
    ☐ My own choice   ☐ Other_____

12. Consider your other professional training experiences which are *not* included in Questions 10 and 11.

In the last <u>five</u> years (starting from July, 1992), did you receive the following professional training? If yes, <u>how useful was it</u>? (If you taught less than 5 years, only consider experience since you began teaching)
(4 = Very useful   3 = Useful   2 = Not very useful   1 = Not useful)

Training focusing on	Received		How Useful			
textbooks and other teaching resources	Yes	No	4	3	2	1
using computer/calculator for teaching math	Yes	No	4	3	2	1
new teaching methods and strategies	Yes	No	4	3	2	1
how to teach particular math topics	Yes	No	4	3	2	1

If you received such training, why did you receive it? (check all that apply)
☐ Required by the State   ☐ Required by the district
☐ My own choice   ☐ Other_____

E. By "professional activities" we **exclude** those <u>specifically</u> designed for professional <u>training</u> which are considered above. We distinguish two kinds of professional activities:

1. Those <u>organized</u> by some organization such as *general* conference, seminar, workshop, etc.
2. Those <u>not organized</u> by an organization, such as informal exchanges with colleagues, reading professional journals and books, etc.

---

13. a. In the last <u>five</u> years, about how many times did you attend professional activities of local, state, and national/regional organizations?

      Local or State   National/Regional
      _____times   _____times

    b. If you attended such activities, why did you attend? (check all that apply)
    ☐ Required by the State   ☐ Required by the district
    ☐ My own choice   ☐ Other_____

c. How useful were they in enhancing your knowledge of *how to teach mathematics?*

	Very useful	Useful	Not very useful	Not useful
Local or State level	4	3	2	1
National/Regional level	4	3	2	1

14. a. During this school year (1996–97), about how often did you attend professional activities organized at or for your school?

	Once per					Other
	Year	Semester	month	2 weeks	week	
School/Departmental level	Year	Semester	month	2 weeks	week	_____
District/County level	Year	Semester	month	2 weeks	week	_____

b. How useful were they in enhancing your knowledge of *how to teach mathematics?*

	Very useful	Useful	Not very useful	Not useful
School/Departmental level	4	3	2	1
District/County level	4	3	2	1

15. a. During this school year, about how often did you do the following?

	Almost daily	2 or 3 times a week	Weekly or biweekly	Monthly	Rarely	Never
Classroom observation	5	4	3	2	1	0
Informal exchange with colleagues	5	4	3	2	1	0
Reading professional journals and books	5	4	3	2	1	0

b. How useful were they in enhancing your knowledge of *how to teach mathematics?*

	Very useful	Useful	Not very useful	Not useful
Classroom observation	4	3	2	1
Informal exchange with colleagues	4	3	2	1
Reading professional journals and books	4	3	2	1

> **F.** Now we would like you to evaluate, based on your experiences, how different sources have contributed to your knowledge of mathematics teaching in different aspects.

16. a. During this school year, how often did you use computers and calculators in the *math* classes you taught?

	Always	Most of the time	Sometimes	Rarely	Never
Calculator	4	3	2	1	0
Computer	4	3	2	1	0

b. How much did the following sources contribute to your knowledge of how to use technology (computer/software, and calculator) for teaching math? (If you did not have some experience, please circle "N/A" correspondingly)

Sources	Very much	Somewhat	Little	No contribution	N/A
a. your experience as a school student	4	3	2	1	
b. pre-service training	4	3	2	1	N/A
c. professional training received since becoming a teacher	4	3	2	1	N/A

(Continued)

Sources	Very much	Somewhat	Little	No contribution	N/A
(Continued)					
d. organized professional activities	4	3	2	1	N/A
e. informal exchanges with colleagues	4	3	2	1	N/A
f. reading professional journals and books	4	3	2	1	N/A
g. your own teaching practices and reflection	4	3	2	1	

17. a. During this school year, how often did you utilize concrete materials in the *math* classes you taught?
    ☐ Always ☐ Most of the time ☐ Sometimes ☐ Rarely ☐ Never
    b. How much did the following sources contribute to your knowledge of how to utilize concrete materials for teaching math?

Sources	Very much	Somewhat	Little	No contribution	N/A
a. your experience as a school student	4	3	2	1	
b. pre-service training	4	3	2	1	N/A
c. professional training received since becoming a teacher	4	3	2	1	N/A
d. organized professional activities	4	3	2	1	N/A
e. informal exchanges with colleagues	4	3	2	1	N/A
f. reading professional journals and books	4	3	2	1	N/A
g. your own teaching practices and reflection	4	3	2	1	

Teacher Questionnaire (Chicago Study)

18. a. Think of the most recent period that you taught. Please fill in the following information.

Course name	Title of the textbook for the course	Years of your using this textbook
_____	_____	_____

    b. How do you feel about your knowledge of this textbook in terms of the textbook's overall characteristics, content arrangement and structure, teaching styles implied, etc.?
    ☐ Not very good ☐ Fairly good ☐ Very good

    c. How much did the following sources contribute to your knowledge of this textbook.

Sources	Very much	Somewhat	Little	No contribution	N/A
a. your experience as a school student	4	3	2	1	
b. pre-service training	4	3	2	1	N/A
c. professional training received since becoming a teacher	4	3	2	1	N/A
d. organized professional activities	4	3	2	1	N/A
e. informal exchanges with colleagues	4	3	2	1	N/A
f. reading professional journals and books	4	3	2	1	N/A
g. your own teaching practices and reflection	4	3	2	1	N/A

19. Considering the general teaching strategies and classroom management models you are using for your teaching, how much did the following sources contribute to your knowledge of those strategies and models?

Sources	Very much	Somewhat	Little	No contribution	N/A
a. your experience as a school student	4	3	2	1	
b. pre-service training	4	3	2	1	N/A
c. professional training received since becoming a teacher	4	3	2	1	N/A
d. organized professional activities	4	3	2	1	N/A
e. informal exchanges with colleagues	4	3	2	1	N/A
f. reading professional journals and books	4	3	2	1	N/A
g. your own teaching practices and reflection	4	3	2	1	N/A

> **G.** In math teaching, there are different ways to represent new math topics (e.g., new concepts, theorems, formulas, and procedures) to students.

For Questions 20 and 21, please recall the **last two** lessons you taught that *contained* **new mathematics topics**, what is the new topic in each lesson (if the lesson had more than one new topic, choose the most important one)? What was the way you represented it to students? How did you get to know the way?

Following is a sample:

The new topic:

SAMPLE: The formula $(a + b)^2 = a^2 + 2ab + b^2$

What was the way you represented it to students?

SAMPLE: I used the following diagram and asked students to find the big square's area and its four smaller rectangles' areas, and hence led them to get the relationship.

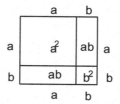

or: I had students use a computer program to evaluate $(a + b)^2$ and $a^2 + 2ab + b^2$ respectively for several values of (a, b), and led them to discover the relationship.

How did you get to know the way?

SAMPLE: I learned this way when I attended a workshop organized by NCTM (or, from observing my colleague's teaching, from informal exchange with a colleague, from reading a magazine, from the textbook, from my own experience/reflection, etc.)

20. Refer to the above description and the sample. Complete the following for the <u>first</u> lesson.
    The new topic:

    What was the way you represented it to students?

    How did you get to know the way?

21. Refer to the above description and the sample. Complete the following for the <u>second</u> lesson.

The new topic:

What was the way you represented it to students?

How did you get to know the way?

22. During this school year, when teaching lessons which included new math topics, how often did you use the following sources to design the way to represent them to students?

Sources being used	Always	Most of the time	Sometimes	Rarely	Never
textbook/teachers' notes	4	3	2	1	0
professional journals or books	4	3	2	1	0
exchanges with colleagues	4	3	2	1	0
your own knowledge	4	3	2	1	0

> Thank you very much. Please place the completed form in the enclosed envelope, and give it sealed to the chair of your math department, from whom we will collect it.

Appendix 1B*

Notes for Classroom Observation

The main purpose of classroom observation in this study is to identify, from the perspective of pedagogy, how teachers actually teach in classrooms, and what kind of pedagogical knowledge teachers actually demonstrate and utilize in their teaching, so specific questions can be raised and asked during the interview about how those teachers developed relevant pedagogical knowledge.

The following questions are intended to guide what observable aspects of the class should be observed in the classroom. The questions should be reviewed carefully before an observation is conducted.

1. What is the topic of the lesson? What are the new mathematics concepts or procedures to be introduced to students?
2. What are the purposes of the lesson?
3. What and how does the teacher use the teaching resources (including textbooks, concrete materials, computer and calculator, etc.) in the class?
4. How does the teacher introduce the topic to students? (e.g., how does he/she represent the new mathematics concepts and/or procedures?)
5. What teaching strategies does the teacher employ for the lesson, and how does he/she organize the classroom? (e.g., does he/she use "cooperative

---

*All the instruments shown in Appendices 1B, 1C, and 1D were also directly used in the Singapore study.

learning" including group work? Does she/he use manipulative and hands-on activities? Does he/she ask students to read the text in the classroom? How does the teacher respond to students' behavior in the classroom? What is the locus of the classroom activity? How is the time spent for different activities? etc.)

Appendix 1C

Script for Interviewing Teachers

First of all, thank you for letting me observe your classes and taking time to talk with me. Based on my observation of your classes, I would like to ask you some questions so I can obtain more detailed and specific information. As this information is important for my study, I'd like to tape this conversation to have an accurate record of your response.

(Note: The following questions are intended to be mainly a guidance for the interview. The actual questions to be asked will depend on the on-site context and the classes observed. Whenever appropriate, all the questions will be related or contextualized to what actually happened in the classes observed.)

1. First of all, what were the purposes of the classes I observed?
   How did you decide the purposes? (based on your understanding of the textbook?, based on teachers' notes? How did you know it?)
2. How did you get your knowledge of the textbooks?
3. In the classes I observed, what concept or knowledge do you think is the most difficult for students to learn? What is the most important to learn? How did you know that?
4. I noticed that you used the way to introduce the new topic. How did you know the way you used? Do you think that using that way will make it easier for students to learn the new topic? How did you get to know?

5. I noticed that you used the following teaching strategies (e.g., cooperative learning) in the classes I observed. How did you get to know those strategies?

   Can you name some other teaching methods that you have used recently but not in the classes I observed? How did you learn about those methods?
6. Did you use computer/software and calculators in your classrooms? (If yes, how often? and mainly for what purposes?)

   How did you learn how to use computer/software and calculators for mathematics teaching, i.e., what are the main sources of your knowledge of that? (ask for in-depth explanations)
7. Did you use concrete materials in your teaching? Can you give me an example to explain how you used them?

   How did you learn how to use concrete materials for teaching mathematics? How did you know that the concrete materials were helpful here?
8. It is said that teachers teach the way they were taught. Do you agree with the statement or not? Can you identify a teacher or more than one teacher when you were at school that significantly influenced the way you teach? (If yes) What are the influences?

   Are there differences between the way you teach and the way you were taught? If yes, what are the differences?
9. Are there major changes of your teaching strategies and teaching styles in recent years? If so, what are the changes? How did you know that?
10. What have been your major sources for you to develop your pedagogical knowledge in your career?
11. How often do you have informal exchanges with your colleagues? Does that help? Can you tell an example that happened last week or this week?
12. It is often said that teachers are too busy to read professional journals and books. What do you think of that statement? How often do you read professional journals and books? Is that helpful? Can you give me an example of knowledge that you obtained from reading professional journals or books?

13. For teachers to seek improving their knowledge of how to teach mathematics more effectively, what do you think are the main difficulties according to your own experiences?

    What do you think needs to be done in order to improve teachers' knowledge of how to teach?

Appendix 1D

Script for Interviewing Math Chairs

First of all, I want to say "thank you" for supporting my survey in your department and taking time to talk with me. Based on the information I collected from your department, I would like to ask you some questions so I can obtain more detailed and specific information. As this information is important for my study, I'd like to tape this conversation to have an accurate record of your response.

(Note: The following questions are intended to be mainly a guidance for the interview. The actual questions to be asked will depend on the context of the school and the department.)

1. How long have you been the chair of the department?

    Since you became the chair, how do you organize professional activity in your department? What are the usual forms? What are the purposes of those professional activities?

2. How often does the department have formal professional activities, such as departmental meeting, seminar, workshop? (Once a week, two weeks, monthly, etc.?)

    What is the main reason that you have this amount of professional activities in the department? (What is the main difficulty? Do teachers have strong motivation and enough time?)

3. Does your department/school have special training programs for new teachers, such as mentoring or apprenticing plans for them?

If yes, what are the main things your department hopes new teachers will learn? How does it fare? Do the new teachers feel that they learned well from those kinds of programs?

If not, how does your department/school help new teachers if they encounter problems or you find they are not capable enough in teaching? Can you give me an example?

4. Does your department have particular plans to further develop experienced teachers' knowledge and ability to teach mathematics?

    If yes, how are they going on now? If not, what does your department/school usually do to support their professional development?

5. It is said that teachers teach the way they were taught. Do you agree with the statement or not? Can you identify a teacher or more than one teacher when you were at school that significantly influenced the way you teach? (If yes) What are the influences?

    Are there differences between the way you teach and the way you were taught? If yes, what are the differences?

6. Are there major changes of your teaching strategies and teaching styles in recent years? If so, what are the changes? How did you know that?

7. What have been your major sources for you to develop your pedagogical knowledge in your career?

8. It is often said that teachers are too busy to read professional journals and books. What do you think of that statement? How often do you read professional journals and books? Is that helpful? Can you give me an example of knowledge that you obtained from reading professional journals or books?

9. For teachers to seek improving their knowledge of how to teach mathematics more effectively, what do you think are the main difficulties according to your own experiences?

    What do you think needs to be done in order to improve teachers' knowledge of how to teach?

Appendix 1E

A Profile of Teacher Participants (Chicago Study)

Note: All the data presented below are based on the Chicagoan teachers' responses to the questionnaire survey.

Table 1E.1 Distribution of teachers by gender and school

	Male	Female	Total
School A	10	16	26
School B	8	14	22
School C	13	8	21
Total	31	38	69

Table 1E.2 Distribution of teachers by age and school

	Age					
	$[20, 30)$[a]	$[30, 40)$	$[40, 50)$	$[50, 60)$	No response	Total
School A	10	5	6	4	1	26
School B	2	8	6	6	0	22
School C	5	3	5	8	0	21
Total	17	16	17	18	1	69

[a] $20 \leq \text{Age} < 30$.

Table 1E.3 Distribution of teachers by highest educational degree and school

	Degree			
	Bachelor's	Master's	Doctorate	Total
School A	7	19	0	26
School B	2	20	0	22
School C	3	17	1	21
Total	12	56	1	69

Table 1E.4 Distribution of the majors and minors of the degrees teachers earned

	Mathematics & Mathematics Education	Education	Others
Major	63 (91.3%)	6 (8.7%)	16 (23.2%)
Major & Minor	65 (94.2%)	4 (5.8%)	39 (56.5%)

Note: Mathematics & Mathematics education — including both mathematics and statistics, curriculum and instruction, and educational psychology in mathematics. Education — including educational administration, secondary education, educational and social policy, etc. Others — including computer, administration, recreation administration, physical education, etc.

Table 1E.5 Distribution of teachers with different lengths of experiences of teaching mathematics

	0–5 years	6–15 years	16 + years	Average years of teaching
School A	7	9	10	13.3
School B	4	4	14	18.5
School C	6	4	11	16.5
Total	17	17	35	15.9

Note: The distribution by the length of teaching any subject is basically the same as that by teaching mathematics. The average length of teaching any subject for all the 69 teachers is 16.2 years, slightly longer than that of teaching mathematics.

Appendix 1F

Main Results of Logistic Regression Analyses (Chicago Study)

Note: This appendix contains six tables (Tables 1F.1 to 1F.6), presenting the main results of the logistic regression analysis on the data collected from the questionnaire survey. An explanation is given for Table 1F.1, which can be similarly applied to Tables 1F.2 to 1F.5.

Pedagogical Curricular Knowledge — Knowledge of Textbooks

Defining p_1 as the probability of choosing "no contribution", p_2 as the probability of choosing "little (contribution)", p_3 as the probability of choosing "somewhat (contribution)", and p_4 as the probability of choosing "very much (contribution)", we can see from the logistic regression,

$$\text{logit}(p_1) = \log\frac{p_1}{1-p_1} = -0.0602 + 0.4551^*X_A + 0.8491^*X_B$$
$$- 0.4284^*X_C + 0^*X_D - 2.5298^*X_E + 0.7346^*X_F$$
$$- 4.0523^*X_G;$$

Table 1F.1 Logistic regression on the data about the contribution of different sources to teachers' knowledge of textbooks

Variable[a]	DF	Parameter Estimate	Standard Error	Wald Chi-Square	Pr > Chi-Square	Standardized Estimate
INTERCP1	1	−0.0602	0.2295	0.0687	0.7932	—
INTERCP2	1	0.9609	0.2360	16.5829	0.0001	—
INTERCP3	1	2.3861	0.2724	76.7085	0.0001	—
A	1	0.4551	0.3289	1.9153	0.1664	0.087895
B	1	0.8491	0.3428	6.1364	0.0132	0.163987
C	1	−0.4284	0.3157	1.8422	0.1747	−0.082744
D[b]	0	0	—	—	—	—
E	1	−2.5298	0.3444	53.9533	0.0001	−0.488581
F	1	0.7346	0.3381	4.7208	0.0298	0.141867
G	1	−4.0523	0.4260	90.4748	0.0001	−0.782612

Note: The score chi-square test for the proportional odds assumption is 14.6452, with DF = 12. The p-value for the test is 0.2614, which indicates the proportional odds assumption is reasonable, and the model decently fits the data.

[a] For the explanatory variables, A = Experience as student, B = Preservice training, C = Inservice training, D = Organized professional activities, E = Informal exchanges with colleagues, F = Reading professional journals and books, G = Own teaching experiences and reflection. For each teacher, seven repeated observations were created for the data process. They are, in terms of the values of explanatory variables, (1,0,0,0,0,0,0,), (0,1,0,0,0,0,0), (0,0,1,0,0,0,0), (0,0,0,0,0,0,0,0), (0,0,0,0,1,0,0,0), (0,0,0,0,0,1,0), and (0,0,0,0,0,0,1). Total number of observations is 68 × 7 = 476, as one teacher did not provide the information. For the response variable Y, there are four response levels: 4 = very much; 3 = somewhat, 2 = little, and 1 = no contribution.
[b] The parameter for D is set to 0, as unsaturated models are used here.

$$\text{logit}(p_1 + p_2) = \log\frac{p_1 + p_2}{1 - p_1 - p_2} = -0.9609 + 0.4551^* X_A$$

$$+ 0.8491^* X_B - 0.4284^* X_C$$

$$+ 0^* X_D - 2.5298^* X_E$$

$$+ 0.7346^* X_F - 4.0523^* X_G;$$

$$\text{logit}(p_1 + p_2 + p_3) = \log\frac{p_1 + p_2 + p_3}{1 - p_1 - p_2 - p_3} = 2.3861 + 0.4551^* X_A$$

Main Results of Logistic Regression Analyses (Chicago Study)

$$+ 0.8491^* X_B - 0.4284^* X_C$$

$$+ 0^* X_D - 2.5298^* X_E$$

$$+ 0.7346^* X_F - 4.0523^* X_G;$$

In the above models, $X_A, \ldots, X_G$ are dummy variables, each being 0 or 1, and $\sum X^2 = 1$.

An example to explain the models is that, using the first model, let $X_A = 1$, $X_B, \ldots, X_G$ being 0, then $\log \frac{p_1}{1-p_1} = -0.0602 + 0.4451 = 0.3849$. Therefore, $p_1 = \frac{e^{0.3849}}{1+e^{0.3849}} = .595$. That is, the logit model predicts, when evaluating the contribution of source A (experience as student), 59.5% teachers chose Y = 1 ("no contribution"); in contrast, let $X_G = 1$, $X_A, \ldots, X_F$ being 0, we can calculate $p_1 = 0.016$. Namely, when evaluating the contribution of Source G (teachers' own teaching experiences and reflection), only 1.6% teachers chose Y = 1 ("no contribution"). Actually, the magnitudes of these slope estimates imply teachers' preferences of choosing Y = 1, relative to the explanatory variable D (note: we set 0 for Source D). Similar explanations apply to the other two models.

For p_4, because $p_4 = 1 - p_1 - p_2 - p_3$, it means that the smaller the coefficient of an explanatory variable is, the bigger p_4 is. In fact, since $\log \frac{p_1+p_2+p_3}{1-p_1-p_2-p_3} = \log \frac{1-p_4}{p_4}$, from the above third model, we can obtain that when evaluating the contribution of Source A (experience as student), only 5.5% teachers chose Y = 4 ("very much contribution"); but when evaluating the contribution of Source G (teachers' own teaching experiences and reflection), as high as 84.1% teachers chose Y = 4 ("very much contribution").

In short, according to the parameter estimates, the order of importance of the sources to the development of teachers' knowledge of textbooks are "teachers' own teaching experiences and reflection" (G: −4.0523), "informal exchanges with colleagues" (E: −2.5298), "inservice training" (C: −0.4284), "organized professional activities" (D: 0), "experience as student" (A: 0.4551), "reading professional journals and books" (F: 0.7346), and "preservice training" (B: 0.8491).

Pedagogical Curricular Knowledge — Knowledge of Technology

Table 1F.2 Logistic regression on the data about the contribution of different sources to teachers' knowledge of technology

Variable[a]	DF	Parameter Estimate	Standard Error	Wald Chi-Square	Pr > Chi-Square	Standardized Estimate
INTERCP1	1	−2.7426	0.2882	90.5371	0.0001	—
INTERCP2	1	−1.3767	0.2563	28.8597	0.0001	—
INTERCP3	1	0.8823	0.2433	13.1464	0.0003	—
A	1	2.7950	0.3565	61.4673	0.0001	0.539793
B	1	2.9424	0.3605	66.6198	0.0001	0.568267
C	1	−0.9082	0.3333	7.4226	0.0064	−0.175398
D[b]	0	0	—	—	—	—
E	1	−1.6493	0.3538	21.7331	0.0001	−0.318527
F	1	1.9272	0.3395	32.2251	0.0001	0.372208
G	1	−1.9138	0.3662	27.3090	0.0001	−0.369616

Note: The score chi-square test for the proportional odds assumption is 7.9216, with DF = 12. The p-value for the test is 0.7912, which indicates the proportional odds assumption is reasonable, and the model fits nicely.

[a] For the explanatory variables, A = Experience as student, B = Preservice training, C = Inservice training, D = Organized professional activities, E = Informal exchanges with colleagues, F = Reading professional journals and books, G = Own teaching experiences and reflection. For each teacher, seven repeated observations were created for the data process. They are, in terms of the values of explanatory variables, (1,0,0,0,0,0,0,), (0,1,0,0,0,0,0), (0,0,1,0,0,0,0), (0,0,0,0,0,0,0,0), (0,0,0,0,1,0,0,0), (0,0,0,0,0,1,0), and (0,0,0,0,0,0,1). Total number of observations is $67 \times 9 = 469$, as two teachers did not provide the information. For the response variable Y, there are four response levels: 4 = very much; 3 = somewhat, 2 = little, and 1 = no contribution.

[b] The parameter for D is set to 0, as unsaturated models are used here.

Pedagogical Curricular Knowledge — Knowledge of Concrete Materials

Table 1F.3 Logistic regression on the data about the contribution of different sources to teachers' knowledge of concrete materials

Variable[a]	DF	Parameter Estimate	Standard Error	Wald Chi-Square	Pr > Chi-Square	Standardized Estimated
INTERCP1	1	−1.6340	0.2414	45.8060	0.0001	—
INTERCP2	1	−0.1157	0.2236	0.2676	0.6049	—
INTERCP3	1	1.8805	0.2523	55.5459	0.0001	—
A	1	1.7641	0.3266	29.1839	0.0001	0.340700
B	1	1.2824	0.3176	16.3061	0.0001	0.247663
C	1	−0.0902	0.3108	0.0842	0.7717	−0.017412
D[b]	0	0	—	—	—	—
E	1	−1.8788	0.3321	32.0077	0.0001	−0.362838
F	1	0.9111	0.3133	8.4563	0.0036	0.175951
G	1	−1.8177	0.3308	30.1918	0.0001	−0.351045

Note: The score chi-square test for the proportional odds assumption is 7.5689, with DF = 12. The p-value for the test is 0.8178, which indicates the proportional odds assumption is reasonable, and the model fits the data nicely.

[a] For the explanatory variables, A = Experience as student, B = Preservice training, C = Inservice training, D = Organized professional activities, E = Informal exchanges with colleagues, F = Reading professional journals and books, G = Own teaching experiences and reflection. For each teacher, seven repeated observations were created for the data process. They are, in terms of the values of explanatory variables, (1,0,0,0,0,0,0,), (0,1,0,0,0,0,0), (0,0,1,0,0,0,0), (0,0,0,0,0,0,0), (0,0,0,0,1,0,0,0), (0,0,0,0,0,1,0), and (0,0,0,0,0,0,1). Total number of observations is 69 × 7 = 483. For the response variable Y, there are four response levels: 4 = very much; 3 = somewhat, 2 = little, and 1 = no contribution.

[b] The parameter for D is set to 0, as unsaturated models are used here.

Pedagogical Instructional Knowledge

Table 1F.4 Logistic regression on the data about the contribution of different sources to teachers' PIK

Variable[a]	DF	Parameter Estimate	Standard Error	Wald Chi-Square	Pr > Chi-Square	Standardized Estimated
INTERCP1	1	−2.1342	0.2564	69.3008	0.0001	—
INTERCP2	1	−0.4099	0.2267	3.2684	0.0706	—
INTERCP3	1	1.5961	0.2483	41.3252	0.0001	—
A	1	0.9635	0.3156	9.3233	0.0023	0.186080
B	1	0.8041	0.3144	6.5401	0.0105	0.155286
C	1	−0.7146	0.3173	5.0721	0.0243	−0.137999
D[b]	0	0	—	—	—	—
E	1	−2.5913	0.3635	50.8325	0.0001	−0.500450
F	1	1.1670	0.3174	13.5152	0.0002	0.225379
G	1	−3.4997	0.4347	64.8189	0.0001	−0.675888

Note: The score chi-square test for the proportional odds assumption is 8.9678, with DF = 12. The p-value for the test is 0.7057, which indicates the proportional odds assumption is reasonable, and the model nicely fits the data.

[a]For the explanatory variables, A = Experience as student, B = Preservice training, C = Inservice training, D = Organized professional activities, E = Informal exchanges with colleagues, F = Reading professional journals and books, G = Own teaching experiences and reflection. For each teacher, seven repeated observations were created for the data process. They are, in terms of the values of explanatory variables, (1,0,0,0,0,0,0,), (0,1,0,0,0,0,0), (0,0,1,0,0,0,0), (0,0,0,0,0,0,0), (0,0,0,0,1,0,0,0), (0,0,0,0,0,1,0), and (0,0,0,0,0,0,1). Total number of observations is 69 × 7 = 483. For the response variable Y, there are four response levels: 4 = very much; 3 = somewhat, 2 = little, and 1 = no contribution.

[b]The parameter for D is set to 0, as unsaturated models are used here.

Knowledge Sources Used to Represent New Topics

Table 1F.5 Logistic regression on the data about the frequencies of different sources teachers used to design the way to represent new mathematics topics

Variable[a]	DF	Parameter Estimate	Standard Error	Wald Chi-Square	Pr > Chi-Square	Standardized Estimate
INTERCP1	1	−5.3184	0.5669	88.0080	0.0001	—
INTERCP2	1	−3.6717	0.4911	55.9087	0.0001	—
INTERCP3	1	−0.3564	0.2350	2.3011	0.1293	—
INTERCP4	1	1.4109	0.2614	29.1420	0.0001	—
A[b]	0	0	—	—	—	—
B	1	3.7316	0.5255	50.4161	0.0001	0.892521
C	1	−0.5422	0.3214	2.8460	0.0916	−0.129688
D	1	−2.4319	0.3746	42.1537	0.0001	−0.581650

Note: The score chi-square test for the proportional odds assumption is 6.4595, with DF = 9. The p-value for the test is 0.6932, which indicates the proportional odds assumption is reasonable, and the model nicely fits the data.
[a]For the explanatory variables, A = textbook/teachers' notes, B = professional journals and books, C = exchanges with colleagues, D = your own knowledge. For the response variable Y, there are five response levels: 4 = always, 3 = most of the time, 2 = sometimes, 1 = rarely, and 0 = never.
[b]The parameter for A is set to 0, as unsaturated models are used here.

Pedagogical Content Knowledge

Table 1F.6 Logistic regression on the data about the contribution of different sources to teachers' PCnK

Variable[a]	DF	Parameter Estimate	Standard Error	Wald Chi-Square	Pr > Chi-Square	Standardized Estimate
INTERCP1	1	−2.0794	0.3062	46.1235	0.0001	—
A	1	−0.7538	0.5198	2.1026	0.1470	−0.137519
B	1	−1.4759	0.6608	4.9892	0.0255	−0.269265
C	1	−0.7538	0.5198	2.1026	0.1470	−0.137519
D[b]	0	0	—	—	—	—

(*Continued*)

Table 1F.6 (*Continued*)

Variable[a]	DF	Parameter Estimate	Standard Error	Wald Chi-Square	Pr > Chi-Square	Standardized Estimate
E	1	1.3863	0.3680	14.1918	0.0002	0.252916
F	1	−0.7538	0.5198	2.1026	0.1470	−0.137519
G	1	2.0794	0.3616	33.0620	0.0001	0.379374
H	1	0.6581	0.3910	2.8328	0.0924	0.120056

Note: The likelihood ratio (−2 Log L) chi-square test statistic is 143.161 with DF = 7, and p = 0.0001, which indicates the combined effects of the explanatory variables (the sources) in the model are significant.

[a] For the explanatory variables, A = Experience as student, B = Preservice training, C = Inservice training, D = Organized professional activities, E = Informal exchanges with colleagues, F = Reading professional journals and books, G = Own teaching experiences and reflection, H = Textbook. The response variable Y is a dummy one, 1 = yes; 0 = no.

[b] The parameter for D is set to 0, as unsaturated models are used here.

Defining p as the probability of choosing "yes", the logistic regression model is

$$\text{logit}(p) = \log\frac{p_1}{1-p_1} = -2.0974 - 0.7538^*X_A - 1.4759^*X_B$$
$$-0.7538^*X_C + 0^*X_D + 1.3683^*X_E - 0.7538^*X_F$$
$$+2.0794^*X_G + 0.6581^*X_H$$

Each parameter, the coefficient of each variable in the model refers to the effect of the variable on the log odds that Y = 1; controlling the other variables, the larger a parameter, the bigger the effect of the variable (the source).

Appendix 2A

A Profile of Teacher Participants (Singapore Study)

Note: All the data presented below are based on the Singaporean teachers' responses to the questionnaire survey.

Table 2A.1 Numbers of full-time and contract teachers in the six secondary schools in Singapore

	School 2A	School 2B	School 2C	School 2D	School 2E	School 2F	Total
Full-time	8	10	9	15	15	7	64
Contract	1	0	0	0	1	0	2
Total	9	10	9	15	16	7	66

Table 2A.2 Numbers of male and female teachers in the six secondary schools in Singapore

	School 2A	School 2B	School 2C	School 2D	School 2E	School 2F	Total
Male	4	5	7	9	4	4	33
Female	5	5	2	6	12	3	33
Total	9	10	9	15	16	7	66

Table 2A.3 Distribution of the teachers by age in the six secondary schools in Singapore

Age	School 2A	School 2B	School 2C	School 2D	School 2E	School 2F	Total
[20, 30)	1	3	5	7	9	4	29
[30, 40)	4	3	1	2	6	1	17
[40, 50)	2	1	3	6	0	2	14
[50, 60)	2	3	0	0	1	0	6
≥ 60	0	0	0	0	0	0	0
Total	9	10	9	15	16	7	66

Table 2A.4 Academic qualifications of the teachers in the six secondary schools in Singapore

	School 2A	School 2B	School 2C	School 2D	School 2E	School 2F	Total
Diploma	6	3	3	3	2	1	18
Bachelor's	5	7	9	8	14	6	49
PGDE	1	2	3	8	12	5	31
Master's	1	0	0	0	1	0	2
Doctorate	0	0	0	0	0	0	0
Others	1	2	0	3	0	1	7

Note: Numbers do not tally with other tables as some of the teachers did not indicate their academic qualifications while some indicated more than one; for example, some of the teachers acquired a Bachelor's degree and proceeded to obtain a PGDE (Postgraduate Diploma in Education). "Others" are contributed mainly by the older teachers who got only "A" Level or a Certificate in Education.

Table 2A.5 Distribution of the teachers according to their years of teaching experience in the six secondary schools in Singapore

	School 2A	School 2B	School 2C	School 2D	School 2E	School 2F	Total
STG1 (<6 years)	2	5	4	7	11	4	33
STG2 (6–15 years)	3	1	2	2	4	1	13
STG3 (>16 years)	4	4	3	6	1	2	20
Total	9	10	9	15	16	7	66

Appendix 2B

Teacher Questionnaire (Singapore Study)

Name (optional)_____ School_____ Date_____

> This is part of a study into how teachers developed their knowledge of how to teach mathematics. The study is an NIE/NTU research project. The purpose of the study is to provide guidance to schools and teacher-training institutions. All responses will be kept strictly confidential; neither schools nor teachers will be identified.
>
> Your response is very important to us. Please respond to all questions as best as you can. As a way to express our thanks, we shall send a brief report of the study, when available, to each participant if he/she wishes. Please check "Yes" or "No" below.
>
> **Do you wish to receive a brief report of the study?** ☐ Yes ☐ No
>
> We sincerely appreciate your cooperation.

Note: please indicate "n/a" below when a question or item is not applicable to you.

A. Your background information.

1. Staff Category: ☐ Full-time qualified teacher   ☐ Contract teacher
2. Gender: ☐ Male   ☐ Female
3. Age: ☐ [20, 30) ☐ [30, 40) ☐ [40, 50) ☐ [50, 60) ☐ ≥ 60

4. Qualification:

Academic Qualifications Earned	Major	Minor(s)
Diploma	_____	_____
Bachelor's	_____	_____
PGDE	_____	_____
Master's	_____	_____
Doctorate	_____	_____
Other (please specify)	_____	

5. Number of years you have taught: Any subjects   ____years
    Mathematics*   ____years

   Excluding computer courses

> **B.** Recall your experience of learning as a student from primary school to junior college level mathematics classes, as well as your experience of teaching as a teacher.

6. Which of the following strategies did you encounter as a student, or use as a teacher?

Teaching strategies	As student being taught (used by your teachers)			As teacher teaching (used by yourself)	
a. review of previous lessons	Yes	No	DR (Don't remember)	Yes	No
b. teacher explains on new topic and students listen	Yes	No	DR	Yes	No
c. classroom discussion	Yes	No	DR	Yes	No
d. small group work	Yes	No	DR	Yes	No

(*Continued*)

Teacher Questionnaire (Singapore Study)

(Continued)

Teaching strategies	As student being taught (used by your teachers)			As teacher teaching (used by yourself)	
e. students' reading the texts	Yes	No	DR	Yes	No
f. use of calculator	Yes	No	DR	Yes	No
g. use of computer	Yes	No	DR	Yes	No
h. use of concrete materials	Yes	No	DR	Yes	No
i. hands-on activities	Yes	No	DR	Yes	No
j. students do project work	Yes	No	DR	Yes	No

C. By "pre-service training" we mean teacher-preparation training (e.g., a teacher education program) you received <u>before the first time</u> you became a qualified teacher.

7. Did you receive pre-service training? ☐ Yes ☐ No
   If "No", please skip this section and go to Question 11.
   If "Yes", the training was for   ☐ teaching mathematics
   ☐ teaching other subjects.

8. How useful were the following courses of your pre-service training in enhancing your knowledge of *how to teach mathematics*?

	Very useful	Useful	Not very useful	Not useful	DNT (Did not take)
• General educational courses	4	3	2	1	DNT
• Pedagogy courses for teaching mathematics	4	3	2	1	DNT
• Teaching practice in schools	4	3	2	1	DNT

9. During your pre-service training, were you taught the following knowledge or skills?

a. how to present new topics orally	Yes	No	DR
			(Don't Remember)
b. how to manage classrooms	Yes	No	DR
c. how to use computers for teaching mathematics	Yes	No	DR
d. how to use calculators for teaching mathematics	Yes	No	DR
e. how to use concrete materials for teaching mathematics	Yes	No	DR
f. *programmed* learning	Yes	No	DR
g. *discovery* learning	Yes	No	DR
h. *cooperative* learning	Yes	No	DR
i. *constructivism* in mathematics learning	Yes	No	DR
j. how to teach specific mathematics topics	Yes	No	DR

10. Overall, how useful was your pre-service training in enhancing your knowledge of *how to teach mathematics?*
    ☐ Very useful   ☐ Useful   ☐ Not very useful   ☐ Not useful

> **D.** Now we turn to in-service training you received <u>after the first time</u> you became a qualified teacher. "In-service training" is *different from* other general "professional activities", such as <u>general</u> conferences, seminars, and classroom observation.

11. Since becoming a qualified teacher, have you enrolled in one or more degree programs in NIE or other higher learning institutions?
    ☐ Yes   ☐ No
    If yes, please list the degree(s) you earned or you are earning. Then evaluate the usefulness of the degree program(s) in enhancing your knowledge of *how to teach mathematics?*
    (4 = Very useful   3 = Useful   2 = Not very useful   1 = Not useful)

Degree & *Major*	**Usefulness**
_____	4 3 2 1
_____	4 3 2 1

Why did you enroll in the degree program(s)? (check all that apply)
☐ Required by the State   ☐ Required by the district
☐ My own choice   ☐ Other_____

12. Since becoming a qualified teacher, have you taken NIE or other higher learning institutions' courses for diploma or certificate but not for an academic degree (which is included in Question 11)?
    ☐ Yes   ☐ No
    If yes, those courses are in the field(s) of _____
    (e.g., general education, mathematics education, computer science, business administration, etc.)
    Why did you take them? (check all that apply)
    ☐ Required by MOE   ☐ Required by your school
    ☐ My own choice   ☐ Other_____
    How useful were they in enhancing your knowledge of *how to teach mathematics?*
    ☐ Very useful   ☐ Useful   ☐ Not very useful   ☐ Not useful

13. Consider your other in-service training experiences which are *not* included in Questions 11 and 12.
    In the last <u>five</u> years (starting from Jan., 1995), did you receive the following in-service training? If yes, <u>how useful was it</u>? (If you taught less than 5 years, only consider experience since you began teaching)
    (4 = Very useful   3 = Useful   2 = Not very useful   1 = Not useful)

Training focusing on	Received	How Useful
• textbooks and other teaching resources	Yes No	4 3 2 1
• using computer/calculator for teaching mathematics	Yes No	4 3 2 1
• new teaching methods and strategies	Yes No	4 3 2 1
• how to teach particular mathematics topics	Yes No	4 3 2 1

If you received such training, why did you receive it? (check all that apply)
☐ Required by MOE ☐ Required by your school
☐ My own choice ☐ Other_____

E. By "professional activities" we **exclude** in-service training which is already considered in Section D. We distinguish two kinds of professional activities:

1. Those <u>organized</u> by some organization such as *general* conference, seminar, etc.
2. Those <u>not organized</u> by an organization, such as informal exchanges with colleagues, reading professional journals and books, etc.

---

14. a. During <u>the school year</u>, how often did you attend professional activities <u>organized</u> at your school?

	On average, ***Once per***					Other
• Departmental level	year	semester	month	2 weeks	week	_____
• School level	year	semester	month	2 weeks	week	_____

b. How useful were they in enhancing your knowledge of *how to teach mathematics?*

	Very useful	Useful	Not very useful	Not useful
• Department level	4	3	2	1
• School level	4	3	2	1

15. a. In the last <u>five</u> years, how many times did you attend professional activities at Cluster/MOE/NIE and regional/international level.
    Cluster/MOE/NIE: _____ times,
    Regional/International: _____ times
    b. If you attended such activities, why did you attend? (check all that apply)

☐ Required by MOE   ☐ Required by your school
☐ My own choice   ☐ Other_____

c. How useful were they in enhancing your knowledge of *how to teach mathematics?*

	Very useful	Useful	Not very useful	Not useful
• Cluster/MOE/NIE	4	3	2	1
• Regional/International	4	3	2	1

16. a. During the school year, how often did you do the following?

	Almost daily	2 or 3 times a week	Weekly/ biweekly	Monthly	Rarely	Never
• Observe other's class	5	4	3	2	1	0
• Informal exchange with colleagues	5	4	3	2	1	0
• Reading professional journals and books	5	4	3	2	1	0

b. How useful were they in enhancing your knowledge of *how to teach mathematics?*

	Very useful	Useful	Not very useful	Not useful
• Observe other's class	4	3	2	1
• Informal exchange with collages	4	3	2	1
• Reading professional journal and books	4	3	2	1

F. Now we would like you to evaluate, based on your experiences, how different sources have contributed to your knowledge of mathematics teaching in different aspects.

17. a. During the school year, how often did you use computers and calculators in the *mathematics* classes you taught?

	Always	Most of the time	Sometimes	Rarely	Never
• Calculator	4	3	2	1	0
• Computer	4	3	2	1	0

b. How much did the following sources contribute to your knowledge of how to <u>use</u> technology (computer/software, and calculator) for teaching mathematics? (If you did not have some experience, please indicate)

Sources	Very much	Somewhat	Little	No contribution
• your experience as a school student	4	3	2	1
• pre-service training	4	3	2	1
• professional training received since becoming a teacher	4	3	2	1
• organized professional activities	4	3	2	1
• informal exchanges with colleagues	4	3	2	1
• reading professional journals and books	4	3	2	1
• your own teaching practices and reflection	4	3	2	1

18. a. During the school year (1999), how often did you utilize concrete materials/physical models in the *mathematics* classes you taught?
 ☐ Always ☐ Most of the time ☐ Sometimes ☐ Rarely ☐ Never
 b. How much did the following sources contribute to your knowledge of how to utilize concrete materials/physical models for teaching mathematics?

Sources	Very much	Somewhat	Little	No contribution
• your experience as a school student	4	3	2	1
• pre-service training	4	3	2	1
• in-service training received since becoming a teacher	4	3	2	1
• organized professional activities	4	3	2	1
• informal exchanges with colleagues	4	3	2	1
• reading professional journals and books	4	3	2	1
• your own teaching practices and reflection	4	3	2	1

19. a. Think of the most recent period (this semester) that you taught. Please fill in the following information.

Course name	Title of the textbook for the course	Years of your using this textbook
_____	_____	_____

b. How do you feel about your knowledge of this textbook in terms of the textbook's overall characteristics, content arrangement and structure, teaching styles implied, etc.?

☐ Not very good ☐ Fairly good ☐ Very good

c. How much did the following sources contribute to your knowledge of this textbook.

Sources	Very much	Somewhat	Little	No contribution
• your experience as a school student	4	3	2	1
• pre-service training	4	3	2	1

(*Continued*)

Sources	(Continued) Very much	Somewhat	Little	No contribution
• in-service training received since becoming a teacher	4	3	2	1
• organized professional activities	4	3	2	1
• informal exchanges with colleagues	4	3	2	1
• reading professional journals and books	4	3	2	1
• your own teaching practices and reflection	4	3	2	1

20. Considering the general teaching strategies and classroom management models you are using for your teaching, how much did the following sources contribute to your knowledge of those strategies and models?

Sources	Very much	Somewhat	Little	No contribution
• your experience as a school student	4	3	2	1
• pre-service training	4	3	2	1
• professional training received since becoming a teacher	4	3	2	1
• organized professional activities	4	3	2	1
• informal exchanges with colleagues	4	3	2	1
• reading professional journals and books	4	3	2	1
• your own teaching practices and reflection	4	3	2	1

Teacher Questionnaire (Singapore Study)

> **G.** In mathematics teaching, there are different ways to represent new mathematics topics (e.g., new concepts, theorems, formulas, and procedures) to students.

For Questions 21 and 22, please recall the **last two lessons** (this semester) you taught that *contained* **new mathematics topics**, what is the new topic in each lesson (if the lesson had more than one new topic, choose the most important one)? What was the way you represented it to students? How did you get to know the way?
Following is an example:
The new topic:
Example: The formula $(a + b)^2 = a^2 + 2ab + b^2$
What was the way you represented it to students?
Example: I used the following diagram and asked students to find the big square's area and its four smaller rectangles' areas, and hence led them to get the relationship.

	a	b	
a	a^2	ab	a
b	ab	b^2	b
	a	b	

or: I had students use a computer program to evaluate $(a + b)^2$ and $a^2 + 2ab + b^2$ respectively for several values of (a, b), and led them to discover the relationship.
How did you get to know the way?
Example: I learned this way when I attended a workshop organized by MOE (or, from observing my colleague's teaching, from informal exchange with a colleague, from reading a magazine, from the textbook, from my own experience/reflection, etc. Please give the specific information as best as you can)

21. Refer to the above description and the example. Complete the following for the <u>first</u> lesson. (Please use a separate sheet of paper if the space is not enough)

The new topic:

What was the way you represented it to students?

How did you get to know the way?

22. Refer to the above description and the example. Complete the following for the <u>second</u> lesson. (Please use a separate sheet of paper if the space is not enough)
    The new topic:

What was the way you represented it to students?

How did you get to know the way?

23. During the school year (1999), when teaching lessons which included new mathematics topics, how often did you use the following sources to design the way to represent them to students?

Sources being used	Always	Most of the time	Sometimes	Rarely	Never
• textbook/ teachers' notes	4	3	2	1	0
• professional journals or books	4	3	2	1	0
• exchanges with colleagues	4	3	2	1	0
• your own knowledge	4	3	2	1	0

Thank you very much. Please place the completed form in the enclosed envelope, and give it sealed to the HOD/Mathematics, from whom we will collect it.

References

Adelman, N. E., Haslam, M. B., & Pringle, B. A. (1996). *The Uses of Time for Teaching and Learning*. Washington, DC: U. S. Department of Education.
Agresti, A. (1990). *Categorical Data Analysis*. New York: John Wiley.
Agresti, A. (1996). *An Introduction to Categorical Data Analysis*. New York: John Wiley.
Aichele, D. B. (Ed.). (1994). *Professional Development for Teachers of Mathematics* (NCTM Yearbook). Reston, VA: National Council of Teachers of Mathematics.
Alexander, P. A., Schallert, D. L., & Hare, V. C. (1991). How researchers in learning and literacy talk about knowledge. *Review of Educational Research, 61*(3), 315–343.
Ayer, A. J. (1956). *The Problem of Knowledge*. London: Macmillan.
Ball, D. L. (1989). *Breaking with Experience in Learning to Teach Mathematics: The Role of a Preservice Methods Course*. East Lansing, MI: National Center for Research on Teaching Education.
Ball, D. L. (1990a). Teaching mathematics for understanding: What do teachers need to know about subject matter knowledge. In M. M. Kennedy (Ed.), *Teaching Academic Subjects to Diverse Learners* (pp. 63–83). New York: Teachers College Press.
Ball, D. L. (1990b). Prospective elementary and secondary teachers' understanding of division. *Journal for Research in Mathematics Education, 21*(2), 132–144.
Ball, D. L. (1991). Research on teaching mathematics: Making subject matter part of the equation. In J. Brophy (Ed.), *Advances in Research on Teaching: Teachers' Knowledge of Subject Matter as it Relates to their Teaching Practices* (Vol. 2, pp. 1–48). Greenwich, CT: JAI Press.
Barnard, H. C. & Lauwerys, J. A. (1963). *A Handbook of British Educational Terms*. London: George G. Harrap & Co. Ltd.
Baturo, A. & Nason, R. (1996). Student teachers' subject knowledge within the domain of area measurement. *Educational Studies in Mathematics, 31*(3), 235–268.
Begle, E. G. (1972). *Teacher Knowledge and Student Achievement in Algebra*. SMSG Reports, No. 9. Stanford: School Mathematics Study Group.
Begle, E. G. (1979). *Critical Variables in Mathematics Education: Finding from a Survey of the Empirical Literature*. Washington, DC: Mathematical Association of America.
Beijaard, D. & Verloop, N. (1996). Assessing teachers' practical knowledge. *Studies in Educational Evaluation, 22*(3), 275–286.

Berdie, D. R., Anderson, J. F., & Niebuhr, M. A. (1986). *Questionnaires: Design and Use.* Metuchen, NJ: Scarecrow Press.

Bidwell, C. E. & Kasarda, J. D. (1975). School district organization and students achievement. *American Sociological Review, 40*(1), 55–70.

Bierhoff, H. (1996). Laying the foundations of numeracy: A comparison of primary school textbooks in Britain, Germany and Switzerland. *Teaching Mathematics and its Application, 15*(4), 141–160.

Bolte, L. A. (1993). *Preservice teachers' content knowledge of functions: Status, organization, and envisioned application.* Doctoral dissertation, University of Missouri–Columbia.

Book, C., Byers, J., & Freeman, D. (1983). Student expectations and teacher education traditions with which we can and cannot live. *Journal of Teacher Education, 36(1)*, 9–13.

Borg, W. R., Gall, J. P., & Gall, M. D. (1993). *Applying Educational Research: A Practical Guide.* New York: Longman.

Borich, G. D. (1992). *Effective Teaching Methods.* New York: Macmillan Publishing Company.

Borko, H. & Putnam, R. (2003). American Educational Research Association 2004 Annual Meeting call for proposals. *Educational Researcher, 32*(4), 34–45.

Boyd, D., Grossman, P., Hammerness, K., Lankford, H., Loeb, S., Ronfeldt, M., & Wyckoff, J. (2013). Recruiting effective math teachers: Evidence from New York city. *American Educational Research Journal, 49*(6), 1008–1047.

Brickhouse, N. W. (1990). Teachers' beliefs about the nature of science and their relationship to classroom practice. *Journal of Teacher Education, 41*(3), 53–62.

Britzman, D. P. (1991). *Practice Makes Practice: A Critical Study of Learning to Teach.* Albany, NY: State University of New York Press.

Bromme, R. (1994). Beyond subject matter: A psychological topology of teachers' professional knowledge. In R. Biehler, R. Scholz, R. Strasser, & B. Winkelmann (Eds.), *Didactics of Mathematics as a Scientific Discipline* (pp. 73–88). Dordrecht, The Netherlands: Kluwer Academic.

Brown, C. A. & Borko, H. (1992). Becoming a mathematics teacher. In D. A. Grouws (Ed.), *Handbook of Research on Mathematics Teaching and Learning* (pp. 209–242). New York: Macmillan Publishing Company.

Brown, R. (1997). *Advanced Mathematics: Precalculus with Discrete Mathematics and Data Analysis.* Evanston, IL: McDougal Little Inc.

Buchmann, M. (1987). Teaching knowledge: the lights that teachers live by. *Oxford Review of Education, 13*(2), 151–164.

Campbell, D. T. (1988). Evolutionary epistemology. In D. T. Campbell, *Methodology and Epistemology for Social Science: Selected Papers* (Edited by E. S. Overman) (pp. 393–434). Chicago: University of Chicago Press.

Carlsen, W. S. (1988). *The effects of teacher subject-matter knowledge on teacher questioning.* Doctoral dissertation, Stanford University.

Carpenter, T. P., Fennema, E., Peterson, P. L., & Carey, D. A. (1988). Teachers' pedagogical content knowledge of students' problem solving in elementary arithmetic. *Journal for Research in Mathematics Education, 19*(5), 385–401.

References

Carpenter, T. P., Fennema, E., & Franke, M. L. (1997). Cognitive guided instruction: A knowledge base for reform in primary mathematics instruction. *The Elementary School Journal, 97*(1), 3–20.

Carter, K. (1990). Teachers' knowledge and learning to teach. In W. R. Houston, M. Haberman, & J. Sikula (Eds.), *Handbook of Research on Teacher Education* (pp. 291–310). New York: Macmillan Publishing Company.

Chisholm, R. M. (1966). *Theory of Knowledge*. Englewood Cliffs, NJ: Prentice-Hall.

Christensen, R. (1997). *Log-Linear Models and Logistic Regression*. New York: Springer.

Clandinin, D. J. & Connelly, F. M. (1987). Teachers personal knowledge: what counts as "personal" in studies of the personal. *Journal of Curriculum Studies, 19*(6), 487–500.

Clandinin, D. J. & Connelly, F. M. (1995). *Teachers' Professional Knowledge Landscapes*. New York: Teachers College Press.

Clandinin, D. J., Davies, A., Hogan, P., & Kennard, B. (1993). *Learning to Teach, Teaching to Learn*. New York: Teachers College Press.

Cochran, K. F., DeRuiter, J. A., & King, R. A. (1993). Pedagogical content knowledge: An integrative model for teacher preparation. *Journal of Teacher Education, 44*(4), 263–272.

Cochran-Smith, M. & Lytle, S. L. (1993). *Inside/Outside: Teacher Research and Knowledge*. New York: Teachers College Press.

Colton, A. B. & Sparks-Langer, G. M. (1993). A conceptual framework to guide the development of teacher reflection and decision-making. *Journal of Teacher Education, 44*(1), 45–54.

Connelly, F. M. & Clandinin, D. J. (1988). *Teachers as Curriculum Planners*. New York: Teachers College Press.

Connelly, F. M., Clandinin, D. J., & He, M. (1996). Teachers' personal practical knowledge on the professional knowledge landscape. *Journal of East China Normal University Educational Science, 52*(2), 5–16 (English translation version).

Cooney, T. J. (1994). Research and teacher education: In search of common ground. *Journal for Research in Mathematics Education, 25*(6), 608–636.

Cooney, T. J. (1999). Conceptualizing teachers' ways of knowing. In D. Tirosh (Ed.), *Forms of Mathematical Knowledge: Learning and Teaching with Understanding* (pp. 163–187). Dordrecht, The Netherlands: Kluwer.

Copeland, W. D. & Doyle, W. (1973). Laboratory skill training and student teacher classroom performance. *Journal of Experimental Education, 42*(1), 16–21.

Corrie, L. (1997). The interaction between teachers' knowledge and skills when managing a troublesome classroom behavior. *Cambridge Journal of Education, 27*(1), 97–105.

Darling-Hammond, L. & Bransford, J. (Eds.). (2005). *Preparing Teachers for a Changing World: What Teachers Should Learn and are Able to do*. San Francisco, CA: Jossey-Bass.

Deng, Z. & Gopinathan, S. (2005). The information technology masterplan. In J. Tan, & P. T. Ng (Eds.), *Shaping Singapore's Future: Thinking School, Learning Nation* (pp. 22–40). Singapore: Pearson.

Dewey, J. & Bentley, A. F. (1949). *Knowing and the Known*. Boston: The Bacon Press.

Donmoyer, R., Imber, M., & Scheurich, J. J. (Eds.). (1995). *The Knowledge Base in Educational Administration*. Albany, NY: State University of New York Press.

Doren, C. V. (1991). *A History of Knowledge: Past, Present, and Future*. New York: Birch Lane Press.

Dreeben, R. (1996). The occupation of teaching and education reform. In K. Wong (Ed.), *Advances in Educational Policy: Rethinking School Reform in Chicago* (Vol. 2, pp. 93–124). Greenwich, CT: JAI Press.

Ebert, C. L. (1994). *An assessment of prospective secondary teachers' pedagogical content knowledge about functions and graphs*. Doctoral dissertation, University of Delaware. Dissertation Abstracts Online Accession No.: AAI9540518.

Edwards, R. & Nicoll, K. (2006). Experience, competence and reflection in the rhetoric of professional development. *British Educational Research Journal, 32*(1), 115–131.

Eisenberg, T. A. (1977). Begle revisited: Teacher knowledge and student achievement in algebra. *Journal for Research in Mathematics Education, 8*(3), 216–222.

Eisenhart, M., Borko, H., Underhill, R., Brown, C., Jones, D., & Agard, P. (1993). Conceptual knowledge falls through the cracks: Complexities of learning to teach mathematics for understanding. *Journal for Research in Mathematics Education, 24*(1), 8–40.

Eisner, E. W. (1997). The promise and perils of alternative forms of data representation. *Educational Researcher, 26*(6), 4–10.

Elbaz, F. (1981). The teacher's "practical knowledge": Report of a case study. *Curriculum Inquiry, 11*(1), 43–72.

Elbaz, F. (1983). *Teacher Thinking: A Study of Practical Knowledge*. London: Croom Helm.

Engelhart, M. D. (1972). *Methods of Educational Research*. Chicago: Rand McNally & Company.

Eraut, M. (1994). *Developing Professional Knowledge and Competence*. London: The Falmer Press.

Even, R. D. (1989). *Prospective secondary mathematics teachers' knowledge and understanding about mathematics functions*. Doctoral dissertation, Michigan State University. Dissertation Abstracts Online Accession No.: AAG8916476.

Even, R. (1993). Subject-matter knowledge and pedagogical content knowledge: Prospective secondary teachers and the function concept. *Journal for Research in Mathematics Education, 24*(2), 94–116.

Even, R. & Tirosh, D. (1995). Subject-matter knowledge and knowledge about students as sources of teacher presentations of the subject-matter. *Educational Studies in Mathematics, 29*(1), 1–20.

Everitt, B. S. (1992). *The Analysis of Contingency Tables*. London: Chapman & Hall.

Fan, L. (1995). A review of the recent development of mathematics education in the United States: The NCTM 73rd annual meeting. *Mathematics Teaching (Shanghai, China), 5*, 1–3; 6, 4–6.

Fan, L. (2002). In-service training in alternative assessment with Singapore mathematics teachers. *The Mathematics Educator, 6*(2), 77–94.

Fan, L., Chen, J., Zhu, Y., Qiu, X., & Hu, Q. (2004). Textbook use within and beyond Chinese mathematics classrooms: A study of 12 secondary schools in Kunming and Fuzhou of China. In L. Fan, N. Y. Wong, J. Cai, & S. Li (Eds.), *How Chinese Learn Mathematics: Perspectives from Insiders* (pp. 228–261). Singapore: World Scientific.

References

Fan, L. & Cheong, C. (2005). The sources of teachers' pedagogical knowledge: The case of Singapore. Paper presented at the American Educational Research Association Annual Meeting, Montreal, Canada.

Fan, L. & Kaeley, G. S. (1998). Textbooks use and teaching strategies: An empirical study. Paper presented at the American Educational Research Association Annual Meeting, San Diego, CA.

Fan, L., Zhu, Y., & Miao, Z. (2013). Textbook research in mathematics education: Development status and directions. *ZDM-International Journal on Mathematics Education, 45*(5), 633–646.

Fang, Z. (1996). A review of research on teacher beliefs and practices. *Educational Research, 38*(1), 47–65.

Feiman-Nemser, S. & Parker, M. B. (1990). Making subject matter part of the conversation in learning to teach. *Journal of Teacher Education, 41*(3), 32–43.

Fennema, E. & Franke, M. L. (1992). Teachers' knowledge and its impact. In D. A. Grouws (Ed.), *Handbook of Research on Mathematics Teaching and Learning* (pp. 147–164). New York: Macmillan Publishing Company.

Fenstermacher, G. D. (1994). The knower and the known: The nature of knowledge in research on teaching. In L. Darling-Hammond (Ed.), *Review of Research in Education* (pp. 3–56). Washington, DC: American Educational Research Association.

Fink, A. (1995). *How to Ask Questions*. Thousand Oaks, CA: Sage Publications.

Fischer, W. L. (2006). Historical topics as indicators for the existence of fundamentals in educational mathematics: An intercultural comparison. In F. K. S. Leung, K.-D. Graf, & F. J. Lopez-Real (Eds.), *Mathematics Education in Different Cultural Traditions: A Comparative Study of East Asia and the West* (pp. 95–110). New York: Springer.

Foerster, P. A. (1998). *Calculus: Concepts and Applications*. Berkeley, CA: Key Curriculum Press.

Foss, D. H, & Kleinsasser, R. C. (1996). Preservice elementary teachers' views of pedagogical and mathematics content knowledge. *Teaching and Teacher Education, 12*(4), 429–442.

Fujii, T. (2001). The changing winds in Japanese mathematics education. *Mathematics Educational Dialogue, 2001*(November). Retrieved 19 June 2002, from http://www.nctm.org/resources/content.aspx?id=1554.

Fuson, K. C. & Briars, D. J. (1990). Using a base-ten blocks learning/teaching approach for first- and second-grade place-value and multidigit addition and subtraction. *Journal for Research in Mathematics Education, 21*(3), 180–206.

Garoutte, M. W. (1980). *Effects of in-service training upon the pedagogical knowledge of inner city teachers*. Doctoral dissertation, University of Missouri-Columbia. Dissertation Abstracts Online Accession No.: AAG8117429.

Gilbert, W., Hirst, L., & Clary, E. (1987). The NCA Workshop's taxonomy of professional knowledge. In D. W. Jones (Ed.), *Professional Knowledge Base: NCATE Approval* (pp. 38–57). Fortieth Annual Report of the North Central Association Teacher Education Workshop. Flagstaff, AZ: University of North Arizona.

Good, C. V. (Ed.). (1945). *Dictionary of Education*. New York: McGraw-Hill.

Good, T. L. & Brophy, J. E. (1994). *Looking in Classrooms*. New York: HarperCollins.

Graeber, A., Tirosh, D., & Glover, R. (1989). Preservice teachers' misconceptions in solving verbal problems in multiplication and division. *Journal for Research in Mathematics Education, 20*(1), 95–102.

Graybeal, S. S. (1988). *A study of instructional suggestions in fifth-grade mathematics and social studies teacher's guides and textbooks.* Doctoral dissertation, University of Chicago.

Green, T. F. (1971). *The Activities of Teaching.* New York: McGraw-Hill.

Greene, M. (1994). Epistemology and education research: The influence of recent approaches to knowledge. In L. Darling-Hammond (Ed.), *Review of Research in Education* (pp. 423–464). Washington, DC: American Educational Research Association.

Griffin, G. A. (1983). Implications of research for staff development programs. *The Elementary School Journal, 83*(4), 414–425.

Griffin, L. (1996). Pedagogical content knowledge for teachers: Integrate everything you know to help students learn. *Journal of Physical Education, Recreation and Dance, 67*(9), 58–61.

Grimmett, P. P. & Mackinnon, A. M. (1992). Craft knowledge and the education of teachers. In G. Grant (Ed.), *Review of Research in Education* (Vol. 18, pp. 385–456). Washington DC: The American Educational Research Association.

Grossman, P. L. (1988). *A study in contrast: sources of pedagogical content knowledge for secondary English.* Doctoral dissertation, Stanford University. Dissertation Abstracts Online Accession No.: AAG8826145.

Grossman, P. L. (1991). *The Making of a Teacher: Teacher Knowledge and Teacher Education.* New York: Teachers College Press.

Grossman, P. L. (1994). Teachers' knowledge. In T. Husén, & T. N. Postlethwaite (Eds.), *International Encyclopedia of Education* (pp. 6117–6122). Oxford: Elsevier Science Ltd.

Grossman, P. L., Wilson, S. M., & Shulman, L. S. (1989). Teachers of substance: Subject matter knowledge for teaching. In M. C. Reynolds (Ed.), *Knowledge Base for the Beginning Teacher* (pp. 23–36). Oxford: Pergamon Press.

Gudmundsdottir, S. (1990). Values in pedagogical knowledge. *Journal of Teacher Education, 41*(3), 44–52.

Heid, M. K. (1997). The technological revolution and the reform of school mathematics. *American Journal of Education, 106*(1), 5–61.

Hiebert, J. & Carpenter, T. P. (1992). Learning and teaching with understanding. In D. A. Grouws (Ed.), *Handbook of Research on Mathematics Teaching and Learning* (pp. 65–97). New York: Macmillan Publishing Company.

Holmes Group. (1986). *Tomorrow's Teachers.* East Lansing, MI: The Holmes Group.

Holmes Group. (1990). *Tomorrow's Schools.* East Lansing, MI: The Holmes Group.

Holmes Group. (1995). *Tomorrow's Schools of Education.* East Lansing, MI: The Holmes Group.

Hood, P. D. & Cates, C. S. (1978). Alternative approaches to analyzing educational dissemination and linkage roles and functions. San Francisco, CA: Far West Laboratory for Educational Research and Development. (ERIC Document No. ED 166810).

Houston, W. R., Haberman, M., & Sikula, J. (Eds.). (1990). *Handbook of Research on Teacher Education.* New York: Macmillan Publishing Company.

Ishii-Kuntz, M. (1994). *Ordinal Log-Linear Models*. Thousands Oaks, CA: Sage.
Jackson, P. W. (1986). *The Practicing of Teaching*. New York: Teachers College Press.
Johnson, D. R. (1982). *Every Minute Counts: Making your Math Class Work*. Palo Alto, CA: Dale Seymour Publications.
Johnson, D. R. (1986). *Making Minutes Count Even More: A Sequel to Every Minute Counts*. Palo Alto, CA: Dale Seymour Publications.
Johnson, M. (1989). Embodied knowledge. *Curriculum Inquiry, 19*(4), 361–377.
Johnston, S. (1992). Images: A way of understanding the practical knowledge of student teachers. *Teaching and Teacher Education, 8*(2), 123–136.
Jones, M. (1997). Trained and untrained secondary school teachers in Barbados: is there a difference in classroom performance? *Educational Research, 39*(2), 175–181.
Jones, M. G. & Vesilind, E. M. (1996). Putting practice into theory: Changes in the organization of preservice teachers' pedagogical knowledge. *American Educational Research Journal, 33*(1), 91–117.
Kaput, J. J. (1992). Technology and mathematics education. In D. A. Grouws (Ed.), *Handbook of Research on Mathematics Teaching and Learning* (pp. 515–556). New York: Macmillan Publishing Company.
Knapp, J. L., McNergney, R. F., Herbert, J. M., & York, H. L. (1990). Should a Master's degree be required by all teachers? *Journal of Teacher Education, 41*(2), 27–37.
Krammer, H. P. M. (1985). The textbooks as classroom context variable. *Teaching & Teacher Education, 1*(4), 273–278.
Labouff, O. (1996). *Teachers' conceptions of understanding mathematics: The challenge of implementing math reform*. Doctoral dissertation, University of California, Los Angeles. Dissertation Abstracts Online Accession No.: AAI9620746.
Ladson-Billings, G. (1995). Toward a theory of culturally relevant pedagogy. *American Educational Research Journal, 32*(3), 465–491.
Laird, J. (1930). *Knowledge, Belief, and Opinion*. New York: The Century Co.
Langrall, C. W., Thornton, C. A., Jones, G. A., & Malone, J. A. (1996). Enhanced pedagogical knowledge and reflective analysis in elementary mathematics education. *Journal of Teacher Education, 47*(4), 271–282.
Lanier, J. E. & Little, J. W. (1986). Research on Teacher Education. In M. Mittrock (Ed.), *Handbook of Research on Teaching* (pp. 527–569). New York: Macmillan.
Lappan, G. & Theule-Lubienski, S. (1994). Training teachers or educating professionals? What are the issues and how are they being resolved? In D. F. Robitaille, D. H. Wheeler, & C. Kieran (Eds.), *Selected Lectures from the 7th International Congress on Mathematical Education* (pp. 249–261). Sainte-Foy, Quebec: Les Presses de L'Universite Laval.
Larson, R. E., Hostetler, R. P., & Edwards, B. H. (1994). *Calculus with Analytic Geometry*. Lexington, MA: Heath.
Larson, R. E., Kanold, T. D., & Stiff, L. (1995a). *Algebra 1: An Integrated Approach*. Lexington, MA: Heath.
Larson, R. E., Kanold, T. D., & Stiff, L. (1995b). *Algebra 2: An Integrated Approach*. Lexington, MA: Heath.
Lawton, D. & Gordon, P. (1993). *Dictionary of Education*. Kent, U.K.: Hodder & Stoughton.

Lee, B. S. (1992). *An investigation of prospective secondary mathematics teachers' understanding of the mathematical limit concept*. Doctoral dissertation, Michigan State University. Dissertation Abstracts Online No.: AAG9233906.

Lee, S. K. (2012). Epilogue. In P. C. Avila, C. Hui, A. Lam, & J. Tan (Eds.), *PISA: Lessons for and from Singapore* (pp. 29–32). Singapore: The Office of Education Research, National Institute of Education, Singapore.

Leinhardt, G. (1990). Capturing craft knowledge in teaching. *Educational Researcher, 19*(2), 18–25.

Leinhardt, G. & Smith, D. A. (1985). Expertise in mathematics instruction: Subject matter knowledge. *Journal of Educational Psychology, 77*(3), 247–271.

Long, J. S. (1997). *Regression Models for Categorical and Limited Dependent Variables*. Thousand Oaks, CA: Sage.

Lortie, D. C. (1975). *Schoolteacher: A Sociological Study*. Chicago: University of Chicago Press.

Love, J. M. (1985). Knowledge transfer and utilization in education. *Review of Research in Education, 12*(1), 337–386.

Machlup, F. (1980). *Knowledge: Its Creation, Distribution, and Economic Significance: Knowledge and Knowledge Production* (Vol. 1). Princeton, NJ: Princeton University Press.

Mao, T. (1971). Where do correct ideas come from? In *Selected Readings from the Works of Mao Tsetung* (pp. 502–504). Peking: Foreign Languages Press.

Marks, R. (1990). Pedagogical content knowledge: From a mathematics case to a modified conception. *Journal of Teacher Education, 41*(3), 3–11.

McConnell, J. W., Brown, S., Usiskin, Z., Senk, S. L., Widerski, T., Anderson, S., Eddins, S., Feldman, C. H., Flanders, J., Hackworth, M., Hirschhorn, D., Polonsky, L., Sachs, L., & Woodward, E. (1996). *UCSMP Algebra*. Glenview, IL: ScottForesman.

McDiarmid, G. W. (1988). The liberal arts: Will more result in better subject matter understanding? *Theory Into Practice, 29*(1), 21–29.

McDiarmid, G. W. (1990). Challenging prospective teacher's beliefs during early field experience: A quixotic undertaking? *Journal of Teacher Education, 41*(3), 11–20.

McGehee, J. J. (1990). *Prospective secondary teachers' knowledge of the function concept*. Doctoral dissertation, University of Texas at Austin. Dissertation Abstracts Online Accession No.: AAG9116926.

McKersie, W. (1996). Reform Chicago's public schools: Philanthropic persistence, 1987–1993. In K. Wong (Ed.), *Advances in Educational Policy: Rethinking School Reform in Chicago* (Vol. 2, pp. 141–158). Greenwich, CT: JAI Press.

McLymont, E. F. & da Costa, J. L. (1998). Cognitive coaching: The vehicle for professional development and teacher collaboration. Paper presented at the American Educational Research Association Annual Meeting, San Diego.

Meredith, A. (1993). Knowledge for teaching mathematics: Some student teachers' views. *Journal of Education for Teaching, 19*(3), 323–338.

Ministry of Education. (1995). *Singapore National Report for TIMSS 1995*. Singapore: The Ministry of Education.

Ministry of Education. (2000). *Mathematics Syllabus (Lower Secondary)*. Singapore: Curriculum Planning Division.

Monroe, P. (Ed.). (1913). *A Cyclopedia of Education* (Vol. 4). New York: Macmillan.

Moore, D. S. & McCabe, G. P. (1993). *Introduction to the Practice of Statistics.* New York: W. H. Freeman & Company.

Moulton, J. (1997). How do teachers use textbooks? A review of the research literature. Technical Paper No. 74, Health and Human Resources Analysis for Africa Project. Retrieved 3 January 2002, from http://www.dec.org/pdf_docs/PNACB240.pdf

Mullens, J. E., Murnane, R. J., & Willett, J. B. (1996). The contribution of training and subject matter knowledge to teaching effectiveness: A multilevel analysis of longitudinal evidence from Belize. *Comparative Education Review, 40*(2), 139–157.

Myers, C. B. & Myers, L. K. (1995). *The Professional Educator: A New Introduction to Teaching and Schools.* Belmont, CA: Wadsworth.

National Center for Education Statistics. (1994). *Characteristics of Stayers, Movers, and Leavers: Results from the Teacher Followup Survey.* NCES 94-337. Washington, DC: U. S. Department of Education.

National Center for Education Statistics. (1997). *Job Satisfaction Among America's Teachers: Effects of Workplace Conditions, Background Characteristics, and Teacher Compensation,* NCES 97-471. Washington, DC: U. S. Department of Education.

National Commission on Excellence in Education. (1983). *A Nation at Risk: The Imperative for Educational Reform.* Washington, DC: U. S. Governmental Printing Office.

National Commission on Teaching and America's Future. (1996). *What Matters Most: Teaching for America's Future.* New York: National Commission on Teaching and America's Future.

National Council of Teachers of Mathematics. (1989a). *Professional Standards for Teaching Mathematics* (draft version). Reston, VA: Author.

National Council of Teachers of Mathematics. (1989b). *Curriculum and Evaluation Standards for School Mathematics.* Reston, VA: Author.

National Council of Teachers of Mathematics. (1991). *Professional Standards for Teaching Mathematics.* Reston, VA: Author.

National Council of Teachers of Mathematics. (1996). *The 74th Annual Meeting Program Book.* Reston, VA: Author.

Neagoy, M. M. M. (1995). *Teachers' pedagogical content knowledge of recursion.* Doctoral dissertation, University of Maryland (College Park). Dissertation Abstracts Online Accession No. AAI9622118.

Orton, R. E. (1993). Two problems with teacher knowledge. In A. Thompson (Ed.), *Philosophy of Education: Proceedings of Philosophy of Education Society 49th Annual Meeting.* Urbana, IL: Philosophy of Education Society.

Pears, D. (1971). *What is Knowledge?* New York: Harper & Row.

Peers, I. S. (1996). *Statistical Analysis for Education and Psychology Researchers.* London: The Falmer Press.

Peressini, A. L., Epp, S. S., Hollowell, K. A., Brown, S., Ellis, W., McConnell, J. W., Sorteberg, J., Thompson, D. R., Aksoy, D., Birky, G. D., McRill, G., & Usiskin, Z. (1992). *UCSMP Precalculus and Discrete Mathematics.* Glenview, IL: ScottForesman.

Polanyi, M. (1966). *The Tacit Dimension.* Garden City, NY: Doubleday & Company, Inc.

Post, T. R., Harel, G. H., Behr, M. J., & Lesh, R. (1991). Intermediate teachers' knowledge of rational number concepts. In E. Fennema, T. P. Carpenter, & S. J. Lamon (Eds.), *Integrating Research on Teaching and Learning Mathematics* (pp. 177–198). Albany, NY: State University of New York Press.

Potter, V. G. (1987). *Philosophy of Knowledge*. New York: Fordham University Press.
Price, J. (1996, April). We've come far, but we still have far to go. NCTM News Bulletin. Reston, VA: National Council of Teachers of Mathematics.
Prichard, H. A. (1950). *Knowledge and Perception: Essays and Lectures*. Oxford: Oxford University Press.
Quinton, A. (1967). Knowledge and belief. In P. Edwards (Ed.), *The Encyclopedia of Philosophy* (Vol. 4, pp. 345–352). New York: Macmillan.
Reynolds, M. C. (Ed.). (1989). *Knowledge Base for the Beginning Teacher*. Oxford, England: Pergamon Press.
Rhoad, R., Milauskas, G., & Whipple, R. (1991). *Geometry: For Enjoyment and Challenge* (New edition). Evanston, IL: McDougal, Littell & Company.
Robitaille, D. F. & Travers, K. J. (1992). International studies of achievement in mathematics. In D. A. Grouws (Ed.), *Handbook of Research on Mathematics Teaching and Learning* (pp. 687–709). New York: Macmillan Publishing Company.
Rowntree, D. (1981). *A Dictionary of Education*. London: Harper & Row.
Rubenstein, R. N., Schultz, J. E., Senk, S. L., Hackworth, M., McConnell, J. W., Viktora, S. S., Aksoy, D., Flanders, J., Kissane, B., and Usiskin, Z. (1992). *Functions, Statistics, and Trigonometry*. Glenview, IL: ScottForseman.
Rubin, L. (1989). The thinking teacher: Cultivating pedagogical intelligence. *Journal of Teacher Education, 40*(6), 31–34.
Russell, B. (1948). *Human Knowledge: Its Scope and Limits*. London: George Allen & Unwin Ltd.
Russell, B. (1959). *The Problems of Philosophy*. London: Oxford University Press. (First published in 1912)
Russell, B. (1992). *Theory of Knowledge: The 1913 Manuscript* (Edited by E. R. Eames). London: Routledge.
Russell, T. & Munby, H. (1991). Reframing: The role of experiencing in developing teachers' professional knowledge. In D. A. Schön (Ed.), *The Reflective Turn: Case Studies in and on Educational Practice*. New York: Teachers College Press.
Ryle, G. (1949). *The Concept of Mind*. London: Hutchinson. (Reprinted by the University of Chicago Press, 1984)
Sadler, P. M., Sonnert, G., Coyle, H. P., Cook-Smith, N., & Miller, J. L. (2013). The influence of teachers' knowledge on student learning in middle school physical science classrooms. *American Educational Research Journal, 50*(5), 1020–1049.
Scholz, J. M. (1995). Professional development for mid-level mathematics. Paper presented at the Annual Meeting of American Educational Research Association, San Francisco, CA.
Schön, D. A. (1983). *The Reflective Practitioner: How Professionals Think in Action*. New York: Basic Books.
Schön, D. A. (1991). *The Reflective Turn: Case Studies in and on Educational Practice*. New York: Teachers College Press.
Schwab, J. J. (1964). The structure of disciplines: Meanings and significance. In G. W. For & L. Pugno (Eds.), *The Structure of Knowledge and the Curriculum*. Chicago: Rand McNalley.
Schwab, J. J. (1968). The concept of the structure of a discipline. In L. J. Hebert & W. Murphy (Eds.), *Structure in the Social Studies* (pp. 43–56). Washington, DC: National Council for the Social Studies.

Shannon, D. M. (1994). The development of preservice teacher knowledge. *Professional Educator*, 17(1), 31–39.

Senk, S. L., Thompson, D. R., Viktora, S. S., Usiskin, Z., Ahbel, N. P., Levin, S., Weinhold, M. L., Rubenstein, R. N., Jackowiak, J. H., Flanders, J., Jakucyn, N., Halvorson, J., & Pillsbury, G. (1996). UCSMP *Advanced Algebra*. Glenview, IL: ScottForesman.

Shope, R. (1983). *The Analysis of Knowing: A Decade of Research*. Princeton, NJ: Princeton University Press.

Shulman, L. (1986a). Those who understand: Knowledge growth in teaching. *Educational Researcher*, 15(2), 4–14.

Shulman, L. (1986b). Paradigms and research programs in the study of teaching: A contemporary perspective. In M. Mittrock (Ed.), *Handbook of Research on Teaching* (pp. 3–36). New York: Macmillan.

Shulman, L. (1987). Knowledge and teaching: Foundations of the new reform. *Harvard Educational Review*, 57(1), 1–22.

Simon, B. (1981). Why no pedagogy in England? In B. Simon & W. Taylor (Eds.), *Education in the Eighties: The Central Issues* (pp. 124–145). London: Batsford Academic & Educational Ltd.

Sobel, M. E. (1995). The analysis of contingency tables. In G. Arminger, C. C. Clogg, & M. E. Sobel (Eds.), *Handbook of Statistical Modeling for the Social and Behavioral Sciences* (pp. 251–310). New York: Plenum Press.

Sosniak, L. A. & Stodolsky, S. S. (1993). Teachers and textbooks: materials use in four fourth-grade classrooms. *The Elementary School Journal*, 93(3), 249–275.

Sowell, E. J. (1989). Effects of manipulative materials in mathematics instruction. *Journal for Research in Mathematics Education*, 20(5), 498–505.

Stevens, C. & Wenner, G. (1996). Elementary preservice teachers' knowledge and beliefs regarding science and mathematics. *School Science and Mathematics*, 96(1), 2–9.

Stodolsky, S. S. (1989). Is teaching really by the textbook? In P. W. Jackson & S. Haroutunian-Gordon (Eds.), *From Socrates to Software: The Teachers as Text and the Text as Teacher: Eighty-ninth Yearbook of the National Society for the Study of Education* (Part I). Chicago: University of Chicago Press.

Suydam, M. N. & Higgins, J. L. (1977). Activity-based learning in elementary school mathematics: Recommendations from research. Columbus, OH: The Ohio State University. (ERIC Document Reproduction Service No. ED144840).

Sztein, A. E. (Ed.). (2010). *The Teacher Development Continuum in the United States and China*. Washington, DC: The National Academies Press.

Talbert, J. E., McLaughlin, M. W., & Rowan, B. (1993). Understanding context effects on secondary school teaching. *Teachers College Record*, 95(1), 45–68.

Tate, W. F. (1994). Diversity, reform, and professional knowledge: The need for multicultural clarity. In D. B. Aichele (Ed.), *Professional Development for Teachers of Mathematics* (NCTM Yearbook, 1994) (pp. 55–66). Reston, VA: National Council of Teachers of Mathematics.

Tatto, M. T., Schwille, J., Senk, S. L., Ingvarson, L., Peck, R., & Rowley, G. (2008). *Teacher Education and Development Study in Mathematics (TEDS-M): Policy, Practice and Readiness to Teach Primary and Second Mathematics. Conceptual Framework*. Amsterdam: IEA.

Thompson, A. G. (1984). The relationship of teachers' conceptions of mathematics and mathematics teaching to instructional practices. *Educational Studies in Mathematics*, 15(2), 105–127.

Tirosh, D. & Graeber, A. O. (1990). Evoking cognitive conflict to explore preservice teachers' thinking about division. *Journal for Research in Mathematics Education, 21*(2), 98–108.

Turner, R. L. (1990). An issue for the 1990s: The efficacy of the required Master's degree. *Journal of Teacher Education, 41*(2), 38–44.

Turner, R. L., Camilli, G., Kroc, R., & Hoover, J. (1986). Policy strategies, teacher salary incentive, and student achievement: An exploratory model. *Educational Researcher,* 15 (Mar.), 5–11.

Ulep, S. A. (2000). Rethinking the Philippine school mathematics curriculum. *Intersection, 1*(1), 2–11.

Usiskin, Z. (2013). Studying textbooks in an information age — A United States perspective. *ZDM-International Journal on Mathematics Education, 45*(5), 713–723.

Waddington, T. S. H. (1995). *Understanding as connected knowledge.* Doctoral Dissertation, University of Chicago.

Wideen, M. F., Mayer-Smith, J. A., & Moon, B. J. (1996). Knowledge, teacher development and change. In I. F. Goodson, & A. Hargreaves (Eds.), *Teachers Professional Lives.* London: The Falmer Press.

Wilson, J. C. (1926). *Statement and Inference* (Vol. 1). Oxford: Oxford University Press.

Wilson, M. R. (1992). *A study of three preservice secondary mathematics teachers' knowledge and beliefs about mathematical functions.* Doctoral dissertation, University of Georgia. Dissertation Abstracts Online Accession No.: AAG9224753.

Wolf, A. (1921). Knowledge. In F. Watson (Ed.), *The Encyclopedia and Dictionary of Education* (Vol. 2, pp. 937–938). London: Sir Isaac Pitman & Sons, Ltd.

Wong, K., Dreeben, R., Lynn, L. E., & Sunderman, G. L. (1997). *Integrated Governance as a Reform Strategy in the Chicago Public Schools.* Chicago, IL: Department of Education, & Irving B. Harris Graduate School of Public Policy Studies, University of Chicago.

Woodward, A. & Elliott, D. L. (1990). Textbook use and teacher professionalism. In D. I. Elliott, & A. Woodward (Eds.), *Textbooks and Schooling in the United States: Eighty-ninth Yearbook of the National Society for the Study of Education* (Part I, pp. 178–193). Chicago, IL: University of Chicago Press.

Zeichner, K. M. (1999). The new scholarship in teacher education. *Educational Researcher, 28*(9), 4–15.

Zeichner, K. M. & Tabachnick, B. R. (1981). Are the effects of university teacher education "washed out" by school experience? *Journal of Teacher Education, 32*(3), 7–11.

Zhang, D. (2009). *My Personal Experience in Mathematics Education: 1938–2008.* Nanjing, China: Jiangsu Education Publishing House.

Zhu, Y. & Fan, L. (2002). Textbook use by Singaporean mathematics teachers at lower secondary school level. In D. Edge & B. H. Yeap (Eds.), *Proceedings of EARCOME-2 & SEACME-9 Conference* (Vol. 2, pp. 194–201). Singapore: Association of Mathematics Educators.

Znaniecki, F. (1965). *The Social Role of the Man of Knowledge.* New York: Octagon Books, Inc.

Author Index

Adelman, N. E., 214
Agresti, A., 79
Aichele, D. B., 29
Alexander, P. A., 14, 15, 41, 61
Anderson, J. F., 59
Ayer, A. J., 37, 38, 61

Ball, D. L., 9, 24, 26, 31
Barnard, H. C., 44
Baturo, A., 24, 27
Begle, E. G., 22, 23, 24, 25, 107, 113
Beijaard, D., 61
Bentley, A. F., 10
Berdie, D. R., 59
Bidwell, C. E., 24
Bierhoff, H., 235
Bolte, L. A., 27
Book, C., 29, 30
Borg, W. R., 65
Borich, G. D., 156, 176
Borko, H., 54, 121, 274
Boyd, D., 274
Bransford, J., 274
Brickhouse, N. W., 22
Britzman, D. P., 54
Bromme, R., 20
Brophy, J. E., 156, 173
Brown, C. A., 54, 121
Brown, R., 134
Buchmann, M., 42, 43
Byers, J., 29

Campbell, D. T., 38
Carlsen, W. S., 22
Carpenter, T. P., 18, 26, 106
Carter, K., 17, 29
Cates, C. S., 15
Cheong, C., 224
Chisholm, R. M., 10, 11
Christensen, R., 79
Clandinin, D. J., 9, 15, 41, 54
Clary, E., 14, 18
Cochran-Smith, M., 14, 50
Cochran, K. F., 18, 19, 119, 227
Colton, A. B., 45, 120
Connelly, F. M., 9, 15, 41, 54
Cooney, T. J., 5, 17, 22, 24, 274
Copeland, W. D., 23
Corrie, L., 41

da Costa, J., 54
Darling-Hammond, L., 274
DeRuiter, J. A., 18, 19, 227
Descartes, R., 10
Dewey, J., 10, 39
Donmoyer, R., 40, 43
Doren, C. V., 13
Doyle, W., 23
Dreeben, R., 4, 214, 215

Ebert, C. L., 27
Edwards, B. H., 88, 136,
Edwards, R., 274
Einstein, A., 41
Eisenberg, T. A., 23

Eisenhart, M., 28
Eisner, E. W., 23
Elbaz, F., 17, 41
Elliot, D. L., 235
Engelhart, M. D., 65
Eraut, M., 14
Even, R. D., 18, 24, 27, 49
Everitt, B. S., 79

Fan, L., 76, 111, 224, 235, 272
Fang, Z., 15
Feiman-Nemser, S., 9, 33
Fennema, E., 5, 18, 20, 45, 155, 163
Fenstermacher, G. D., 11, 12, 15, 17, 29, 42, 50
Fink, A., 59
Fischer, W. L., 225, 226
Foerster, P. A., 86, 125
Foss, D. H., 32, 52
Franke, M. L., 5, 18, 20, 45, 155, 163
Freeman, D., 29
Fujii, T., 235

Gall, J. P., 65
Gall, M. D., 65
Garoutte, M. W., 33, 53
Gilbert, W., 14, 18, 21, 43
Glover, R., 26
Good, C. V., 44
Good, T. L., 156, 173
Gordon, P., 44
Graeber, A., 26
Graybeal, S. S., 76
Green, T. F., 11
Greene, M., 41
Griffin, G. A., 35
Griffin, L., 18
Grimmett, P. P., 14
Grossman, P. L., 5, 18, 19, 22, 24, 30, 49, 52, 119, 120
Gudmundsdottir, S., 22

Haberman, M., 29
Hare, V. C., 14
Haslam, M. B., 214
He, M., 41

Heid, M. K., 93, 240
Hiebert, J., 106
Hirst, L., 14, 18
Hood, P. D., 15
Hostetler, R. P., 88, 136
Houston, W. R., 29

Imber, M., 40
Ishii-Kuntz, M., 79

Jackson, P. W., 11, 52
Johnson, D. R., 13, 166, 227
Johnson, M., 41
Johnston, S., 5, 43
Jones, M. G., 31, 32, 34, 45, 227

Kaeley, G. S., 76
Kanold, T. D., 87, 131, 133, 140
Kant, I., 10
Kaput, J. J., 93
Kasarda, J. D., 24
King, R. A., 18, 19
Kleinsasser, R. C., 32, 52
Krammer, H. P. M., 76, 235

Labouff, O., 28
Ladson-Billings, G., 21
Laird, J., 10, 11, 39
Langrall, C. W., 32
Lanier, J. E., 34
Lappan, G., ix, 9, 20, 21, 31
Larson, R. E., 87, 88, 131, 133, 136, 140
Lauwerys, J. A., 44
Lawton, D., 44
Lee, B. S., 26
Lee, S. K., 224
Leibniz, G., 41
Leinhardt, G., 14, 19, 22, 24, 25
Little, J. W., 34
Liu, J., xxi
Long, J. S., 79
Lortie, D. C., 5, 29, 30, 51, 199
Love, J. M., 14, 15
Lytle, S. L., 14, 50

Machlup, F., 14, 61
Mackinnon, A. M., 14

Mao, T., 42
Marks, R., 18, 227
Mayer-Smith, J. A., 15
McCabe, G. P., 83
McConnell, J. W., 87, 144
McDiarmid, G. W., 24, 30, 31, 52
McGehee, J. J., 27
McKersie, W., 3
McLaughlin, M., 21
McLymont, E. F, 54
Meredith, A., 18
Miao, Z., 235
Milauskas, G., 84, 89, 132, 138, 141
Monroe, P., 16, 44
Moore, D. S., 83
Moulton, J., 239
Mullens, J. E., 23
Munby, H., 215
Murnane, R. J., 23
Myers, C. B., 156
Myers, L. K., 156

Nason, R., 24, 27
Neagoy, M. M. M., 18
Newton, I., 14
Nicoll, K., 274
Niebuhr, M. A., 59

Orton, R. E., 14, 40, 43, 61

Parker, M. B., 9, 33
Pears, D., 12
Peers, I. S., 83
Peressini, A. L., 87, 144, 145
Polanyi, M., 14, 61
Post, T. R., 26
Potter, V. G., 10, 12
Price, J., 4
Prichard, H. A., 10
Pringle, B. A., 214
Putnam, R., 274

Quinton, A., 10, 11, 40

Reynolds, M. C., 29
Rhoad, R., 84, 88, 89, 132, 138, 141, 149
Robitaille, D. F., 76

Rowan, B., 21
Rowntree, D., 44
Rubenstein, R. N., 147, 149
Rubin, L., 43
Russell, B., 10, 12, 37
Russell, T., 215
Ryle, G., 13, 61

Sadler, P. M., 274
Schallert, D. L., 14
Scheurich, J. J., 40
Scholz, J. M., 33, 34, 53
Schön, D. A., 32, 54, 215
Schwab, J. J., 19
Senk, S. L., 147
Shannon, D. M., 32
Shope, R., 11
Shulman, L., 4, 9, 17–21, 24, 49, 119–120, 130, 137–138, 155
Sikula, J., 29
Simon, B., 44
Smith, D. A., 19, 22, 24, 25
Sobel, M. E., 79
Sosniak, L. A., 76
Sparks-Langer, G. M., 45, 120
Stevens, C., 28
Stiff, L., 87, 131, 133, 140
Stodolsky, S. S., 76
Sztein, A. E., 274

Tabachnick, B. R., 31
Talbert, J. E., 21
Tate, W. F., 24
Tatto, M. T., 274
Theule-Lubienski, S., ix, 9, 20, 21, 31
Thompson, A. G., 4, 24, 27
Tirosh, D., 26, 49
Travers, K. J., 76
Turner, R. L., 24, 25

Ulep, S. A., 235
Usiskin, Z., 235

Verloop, N., 61
Vesilind, E. M., 31, 32, 45, 227

Waddington, T. S. H., 39

Wenner, G., 28
Whipple, R., 84, 89, 132, 138,141
Wideen, M. F., 15
Willett, J. B., 23
Wilson, J. C., 10
Wilson, M. R., 27
Wilson, S. M., 19, 24
Wolf, A., 11

Wong, K., 3
Woodward, A., 235

Zeichner, K. M., 31, 274
Zhang, D., 22
Zhao, D., xxii
Zhu, Y., 235
Znaniecki, F., 16

Subject Index

advanced diploma, 229
Alaska, 49
American Educational Research Association, 4, 18, 224, 274
ANOVA, 124
Aristotle, 10
Australia, 27

Barbados, 34
belief, 11, 14–16, 19, 21, 23, 28, 30–32, 39, 40, 43, 119, 145, 166, 171, 172, 174, 226, 256, 274
    evidentially supported belief, 11
    justified true belief, 11, 40
    objectively grounded belief, 40
Belize, 23
Boston, 111

Cambridge General Certificate of Education "Ordinary" Level Examination (GCE, "O" Level), 222
Canada, 224
centralized institute, 223
chi-square test, 70, 80–83, 96–99, 110–112, 113, 125, 127–129, 159, 161–163, 191, 196, 197, 199, 210, 234, 238, 243, 247, 251, 252, 255, 257, 262, 265, 298, 300–304
Chicago
    Chicago public school, 3
Chicago study, the xix–xxii, 3–7, 9, 37, 75, 209, 221, 225, 226, 228, 231–234, 236, 257, 259, 261, 266, 268, 269, 271–273
Chicago Sun-Times, 58
Chicago Tribune, 58
China, xxi, xxii, 52, 85, 216, 217, 235
Chinese, 221
classroom discussion, 186, 187, 253, 276, 308
classroom instruction, 25
classroom management, 17, 18, 155, 156, 160, 191, 193, 201, 255, 283, 316
classroom observation, xx, 27, 53, 58, 59, 61–63, 65–70, 72, 73, 75, 121, 130, 138, 152, 156, 163, 176, 183, 198, 199, 210, 211, 229–233, 234, 248, 255, 257, 278, 280, 281, 287, 310
classroom teaching, xix, 11, 16, 22, 24, 34, 61, 93, 107, 120, 160, 166, 178, 202, 230, 232, 235, 240, 272
clinical experience, 32
Cluster Superintendent, 229
comparative study, 217, 223, 274
compulsory primary education, 222
conceptual framework, xx, 6, 10, 37, 57, 60, 69, 71, 72, 75, 90, 164, 221, 226, 231
concrete materials, 45, 62, 75, 84, 106–118, 164, 186, 187, 211, 212, 227, 231, 235, 245–248, 253, 261, 262, 264, 268, 269, 276, 277, 282, 287, 290, 301, 309, 310, 314

cooperative learning, 48, 52, 62, 70, 156, 163–170, 172–181, 186–190, 193, 202, 206, 227, 277, 290, 310
curriculum and instruction, 9
curriculum development, 18
curriculum-centralized countries, 52
curriculum-decentralized countries, 52

dichotomy, 13
differentiation, 249
diploma courses, 223

East China Normal University Press, xxi
education, 193
    education system, 224
    educational contexts, 261
    educational policy, 9, 25
    educational reform, 3
    educational reformers, 3
effective teaching, 17, 156
epistemological background, xx, 6, 10, 36
epistemologists, 10, 11, 37, 44
epistemology, 11, 12, 24, 37, 38, 40
Eurasians, 222
experience as learners, 50, 60, 115
Express Course, 222

GCE "Advanced" Level Examination ("A" Level), 223
GCE "Normal" Level Examination ("N" Level), 222, 223
GCE "O" Level Examination, 222, 223, 230
Germany, 235
gross domestic product (GDP), 222
group discussion, 11, 48

Hawaii, 49
heads of departments (HODs), 232, 233, 244, 256, 259, 319
Holmes Group, 3

Illinois, 57, 58, 86
Illinois Council of Teachers of Mathematics (ICTM), 102, 180

Illinois Goal Assessment Program (IGAP), 57, 58, 64, 72, 210
in-service training, 315, 316
Indians, 222
information technology (IT), 240
inservice teacher education, 34
inservice training, 29, 33, 35, 36, 52, 62, 69, 78, 80–82, 86, 92, 94–98, 101, 102, 105–112, 115–117, 125–128, 130, 151–153, 157–161, 167, 176, 178, 179, 181–183, 185, 192–194, 201–203, 205–207, 211–213, 216, 229, 234, 236–251, 253–258, 262–270, 272, 273, 298–302, 304
Institutes of Technical Education (ITE), 223
instructional materials, 45, 48, 55, 75, 210, 235, 261
instruments, 6
intellectual inquiry, 17
International Association for Evaluation of Educational Achievement (IEA), 274
international comparisons, 217
International Congress on Mathematical Education, 20
International Monetary Fund, 222

Japan, 52, 235
junior colleges, 223
justification, 11, 43

knowledge
    conceptual knowledge, 13
    content knowledge, 17
    curricular knowledge, 17
    curriculum knowledge, 18
    direct knowledge, 14
    general knowledge, 14
    indirect knowledge, 14
    knowledge by acquaintance, 14
    knowledge by description, 14
    mathematical knowledge, 274
    nature of knowledge, 10, 37
    practical knowledge, 14, 17, 41
    prior knowledge, 32, 50

Subject Index

procedural knowledge, 15
professional knowledge, xxii, 14, 15, 17, 18, 20, 30, 45, 245, 249, 257, 266, 271, 274
propositional knowledge, 11–13, 23
situated knowledge, 14, 53
subject matter knowledge, 16, 17
tacit knowledge, 16, 53
the knower, 14, 16, 39–42, 50, 54, 226
the knowing, 40, 54
the known, 14, 16, 40–42, 54
theoretical knowledge, 14
theory of knowledge, 39
knowledge base for teaching, 17
knowledge of pedagogy, 5
knowledge of textbooks, 77, 86, 240

length of teaching experience, 258
lifelong learners, 269, 271
Linear Equations, 140
log-linear regression, 79, 96, 109, 127, 158, 201, 210
logistic regression, 95, 234, 237, 241–243, 247, 249, 250, 254, 255, 257, 262, 265, 266, 297, 298, 300–303

Malays, 221
math chairs, 58, 62, 63, 65, 72, 168, 188, 201, 204, 205, 210
mathematical discourse, 46, 49
mathematical tasks, 45
Mathematics, 193
mathematics classes, 245
mathematics education, 4, 28, 48, 54, 68, 111, 193, 274, 296, 311
mathematics pedagogy courses, 52
mathematics syllabus, 249
Mathematics Teacher, 199
mathematics teaching, 4, 27, 31, 32, 34, 53, 59, 88, 111, 173, 193, 237, 241, 245, 248, 253, 281, 290, 313, 317
matrix, 249
medium of instruction, 222
mensuration, 249
mental result, 226

mentoring programs, 204, 205, 214
Metropolitan Mathematics Club of Chicago (MMC), 114
Michigan State University, 9, 29
Ministry of Education (MOE), 222, 224, 229, 230, 240, 256, 257, 311–313, 317
Montreal, 224
Mother Tongue, 222

National Census of Population, 221
National Commission on Excellence in Education, 3
National Commission on Teaching and America's Future, 3
National Council of Teachers of Mathematics (NCTM), 4, 21, 45, 48, 54, 111, 121, 124, 172, 174, 199, 226, 235, 285
National Institute of Education (NIE), 224, 229, 307, 310–313
Netherlands, 235
non-organized professional activities, 53, 198–200
Normal Course, 222, 223
  Normal Academic Course, 222
  Normal Technical Course, 222

official languages, 222
organized professional activities, 174, 176, 179, 181

pedagogical knowledge, xix–xxii, 5–7, 16, 17, 19–23, 25, 29, 31–33, 36, 37, 44, 45, 49–55, 57, 59–61, 63, 67, 69–73, 75, 130, 155, 172, 185–191, 193–195, 197, 200–207, 209–217, 224–228, 231–235, 257–259, 261, 262, 268, 269, 271–275, 287, 290, 294
  pedagogical content knowledge (PCnK), xx, 6, 18–21, 25–27, 30, 32, 33, 45, 47–49, 52, 53, 60, 62, 69, 75, 119–121, 123–131, 133–135, 137–139, 141, 147, 149, 151–153, 185, 186, 194, 210–213, 215, 227,

234, 235, 248–253, 257, 258, 261, 264–266, 269–271, 273, 303
pedagogical curricular knowledge (PCrK), xx, 6, 48, 52, 53, 60, 62, 69, 75, 93, 116–118, 127, 185, 194, 210, 211, 213, 227, 234, 235, 258, 261–264, 268–270, 297, 300, 301
pedagogical instructional knowledge (PIK), xx, 6, 48, 53, 55, 60, 62, 69, 75, 155–161, 175, 181–185, 194, 210–213, 227, 234, 235, 252–256, 258, 261, 266–270, 272, 302
pedagogical training, 29
pedagogy, 5, 9
Philippines, 235
pilot test, 61, 63, 64
Plato, 10
policy makers, 3, 5, 223, 225, 272
educational policy makers, xxii, 258, 271–274
polytechnics, 223
post-secondary education, 223
Postgraduate Diploma in Education (PGDE), 308
practice teaching, 28, 52
pre-university course, 223
preservice, 34
preservice teacher education, 274
preservice trainee, 271
preservice training, xxi, 30, 31, 34–36, 51, 52, 54, 55, 60, 69, 70, 78–82, 85, 87, 92, 94–101, 106–112, 115–118, 125–128, 130, 145, 151, 153, 157–161, 165–168, 174–177, 181–183, 185, 188–193, 201–206, 210–213, 216, 228, 229, 234, 236–242, 244–259, 262–273, 277, 278, 281–284, 298–302, 304, 314–316
Primary School Leaving Examination (PSLE), 222
prior knowledge, 49
problem solving, 257

Professional Standards for Teaching Mathematics, 45, 46
NCTM *Standards*, 55, 92, 107, 120, 155, 240
professional training, 52, 53, 68, 86, 102, 143, 193–195, 214, 244, 278, 279, 281, 282, 314, 316
IT-based professional training, 244
university-based professional training, 192
Programme for International Student Assessment (PISA), 223
project work, 253, 309
project-based learning, 173
psychology, 9, 43, 143
educational psychology, 30, 193, 296
purchasing-power-parity per capita, 222
Pythagorean theorem, 122, 123

qualitative methods, 70, 73, 210, 211, 234, 257

Random House Unabridged Dictionary, 44
Remainder Theorem, 134
research design, xx, 6

San Diego, 125
school administrators, 7, 274
school population, 230
school principals, 259
school ranking, 230
school reform movement, 3
secondary education, 222, 223
self-reflection, 34, 53, 54
senior teachers, 269
Shanghai, xxi, 49
simultaneous equations, 249
Singapore, 235
Singapore classrooms, 253
Singapore Department of Statistics, 222
Singapore study, the, xix, xxi, xxii, 221, 224–228, 232, 233, 235, 257–259, 261, 266, 269, 271, 287, 305, 307
Singapore teachers, 224

Subject Index

social expectation, 25
Special Course, 222
standardized test, 24
Stanford University, 9
statistical analysis, 269, 271
stratified random sample, xx, xxi, 65, 72, 210, 230, 231, 257
streaming policy, 222
structured interviews, 62
student teaching, 31, 32, 52, 101, 145, 165, 173, 188, 190–192, 204, 206, 213, 214, 277
students' learning, 3, 24, 227
studies
  analytical studies, 16
  empirical studies, 16, 22
subject matter knowledge, 19
summer course, 52

teacher attrition, 271
teacher education, xix, 5, 9, 17, 22, 28, 30–32, 34, 49, 71, 111, 213, 225, 258, 271, 272, 274, 309
Teacher Education and Development Study in Mathematics or TEDS-M, 274
teacher educators, 7
teacher improvement, 3, 4
teacher knowledge, xx, 4, 6, 9, 10, 12, 15, 17, 19, 23, 36, 41–43, 49, 53, 209
teacher learning, 274
teacher professional development, xix, xxi, 29, 71, 224, 258
teacher training, 24, 34, 189, 224
teacher-preparation, 277
teacher-training college, 50
teacher-training institutions, 275
teachers
  English teachers, 30, 169
  experienced teachers, 33, 294
  inservice teachers, 33, 229, 272
  junior teachers, 111, 116, 252
  mathematics teachers, xx–xxii, 5, 20, 21, 25, 26, 28, 31, 48, 58, 61, 63–65, 71, 72, 76, 102, 130, 164, 210, 214, 216, 224–226, 230, 233, 235, 249, 253, 257, 259, 274
  mathematics teachers' classroom teaching, 63
  mentor teachers, 33
  new teachers, 204, 214, 271, 293, 294
  novice teachers, 25, 33, 60
  preservice teachers, 26–28, 30–32, 191, 263
  prospective teachers, 26, 30, 33, 34, 52, 60, 111, 190, 214, 258
  science teachers, 34, 169
  senior teachers, 85, 129, 211, 213, 258, 268–270
  untrained teachers, 34
teachers' evaluation, 70, 80, 96, 100, 111, 190, 192, 194, 196, 197, 238, 246, 248, 265, 266
teachers' knowledge, xix, xxii, 3, 5, 9, 12–14, 20–24, 29, 33, 35, 37, 41, 42, 44, 45, 47, 48, 50, 53, 55, 61, 63, 76, 78, 79, 84, 92, 93, 95, 97, 100, 102, 105–111, 115–117, 123, 130, 155, 196, 211, 213, 216, 225, 227, 236, 237, 239–245, 247, 248, 257, 261, 263, 268, 272, 274, 291, 294, 298–301
  conceptual knowledge, 28
  content knowledge, 17, 18, 20, 32
  core knowledge, 36
  curricular knowledge, 47, 262
  curriculum knowledge, 26
  general knowledge, 23, 24, 27, 86–89
  knowledge of context, 19
  knowledge of learners, 18, 20
  knowledge of textbooks, 77, 78, 80, 81, 83, 86, 89–92, 94, 96, 118, 186, 211, 215, 235–240
  lesson structure knowledge, 19
  mathematical knowledge, 11, 40, 46
  subject matter knowledge, 16, 17, 19, 20, 22–27, 31, 33, 36, 45, 48, 49, 120
  substantive knowledge, 19, 27
  syntactical knowledge, 19
  teaching knowledge, 198

teachers' own reflection, 54
teachers' pedagogical knowledge, xx, 69, 70, 209
teachers' professional development, 5, 35, 59, 213, 214, 225, 251, 257, 259, 273
teaching knowledge, 42
teaching methods, 11, 18, 40, 44, 45, 47, 54, 194, 195, 206, 279, 290, 311
The International Monetary Fund's World Economic Outlook Database, 222
The Oxford English Dictionary, 44
Third International Mathematics and Science Study, 223
transformation, 249
Trends in International Mathematics and Science Study (TIMSS), 223, 224, 272

University of Chicago School Mathematics Project (UCSMP), 76, 87, 91, 144, 147
University of Toronto, 9
USA, 235

Virginia, 25

Webster's New Universal Unabridged Dictionary, 44
West Indies, 34
What Matters Most: Teaching for America's Future, 3
working language, 222